# WINGED DEFENSE

Fire Ant Books

General Mitchell
1925

# WINGED DEFENSE

*The Development and Possibilities of Modern Air Power—Economic and Military*

William "Billy" Mitchell

The University of Alabama Press • *Tuscaloosa, AL*

The University of Alabama Press
Tuscaloosa, Alabama 35487-0380

Manufactured in the United States of America

∞

The paper on which this book is printed meets the minimum requirements of American National Standard for Information Sciences-Permanence of Paper for Printed Library Materials, ANSI Z39.48-1984.

Cataloging-in-Publication Data available from the Library of Congress
ISBN 978-0-8173-5605-7

## A NOTE ON BILLY MITCHELL AND HIS BOOK

William "Billy" Mitchell was and continues to be one of the most controversial airmen in our history. Alternately lionized and vilified by military officers and civilian academics, Mitchell was undeniably one of the pivotal figures in the development of American air power. Although he began his 27-year military career in the infantry, serving in the Spanish-American War and along the turbulent Mexican border, he took to airplanes, and to the air, with great zeal, learning to fly in 1916 on his off-duty time. Upon America's entry into the First World War, he was the first U.S. officer to participate in a ground attack with the French in 1917, the first to fly an aircraft over German lines, and the first to earn the War Cross for combat duty.

General Pershing, Commander in Chief of American forces in France, knew Mitchell's talents and placed him in charge of Air Service forces, culminating with Mitchell's command of the combined French, British, Italian, and American air assets that for the first time operated with significant battlefield effect under the principle of centralized control during the Meuse-Argonne offensive in the war's closing months. Mitchell's experiences as the senior American airman in France made a deep impression on him because he saw not only the carnage of the war firsthand but also, he believed, a way to avoid it in future armed conflicts.

Always the activist, Mitchell pushed hard for an independent air force and a Department of Aeronautics within the federal government to give the United States a global lead in the development of air power—in short, to make it an "air-going nation." As Assistant Chief of the Air Service and Director of Military Aviation from 1919 to 1925, Mitchell argued that air power had eclipsed other kinds of military power and would win future wars by attacking an enemy's "vital centers" (industrial centers), breaking both the adversary's ability and will to resist. Although this was the vital core of his message, Mitchell took more direct aim at the Navy, contending that aircraft had made naval vessels obsolete and incapable of defending the United States from invasion. His bombing trials against the *Ostfrieland, New Jersey,* and several other German and obsolete American ships in 1923, which sank both of the capital ships involved and several others, electrified the public and, in the process, made Naval officers his mortal enemies.

The increasingly acrimonious debate between Mitchell, the Department of the Navy, and several other government agencies following the bombing trials led ultimately to Mitchell's removal as Assistant Chief of Air Service. When he then attacked what he viewed as "the incompetency, criminal negligence, and almost treasonable administration of the National Defense by the Navy and War Departments," his court-martial followed in October-December 1925. Found guilty on several charges, including insubordination and conduct unbecoming an officer, Mitchell resigned from the Army Air Service. Yet this was by no means the end of his public advocacy of air power.

Mitchell had already published *Winged Defense*

before his court-martial, and he continued to speak, write, and travel with an almost frenetic energy to share his ideas with the American public and influential members of government and industry. By the time he died on February 17, 1936, at the age of 57, Mitchell had convinced many Americans that both military and civil aviation would be vital to their country's future fortunes. He proved right on both counts.

While Mitchell got several things wrong, particularly regarding his belief in the decisive role of air power as a war-winning instrument in its own right, he got a great deal more right. Air power did prove to be an indispensable contributor to Allied victory in the Second World War and has remained a vital guardian of America's national security since the Air Force achieved independence in 1947. Mitchell's views of civil aviation's importance also proved prescient as the United States became—and remains—the world's premier air power nation. When viewed in this light, Mitchell's theories and ideas about military and civil aviation were for the most part accurate. Yet the legacy of his court-martial presents us with a dilemma: Did Mitchell's methods justify the results they ultimately produced?

In recent years, this most fundamental aspect of the controversy still surrounding Billy Mitchell has led to a certain degree of revisionist thinking, particularly in the Air Force. Serving officers, unlike most of their predecessors at the Air Corps Tactical School in the 1920s and 1930s or succeeding generations up to and including the fighter generals of the 1990s, have come increasingly to question whether Mitchell's means justified the ends he achieved and if the kinds of excesses in which he engaged are ever acceptable behavior for a military officer. Clearly, the answer must be "No." However, they also recognize that there might have been a third way: more careful and diplomatically savvy activism within the structures—and strictures—of the federal government and the military departments. This was the tragedy of William Mitchell: His very temperament and maverick style set him on what now appears to be an unavoidable collision course with his foes. Indeed, it appears as though he was fully prepared to bring the court-martial upon himself and to suffer the consequences. One can only wonder what he might have accomplished had he approached the problem in a more professional and careful fashion.

Whatever else one might say about Mitchell, his ideas made him one of the most significant air power theorists of the interwar period—in fact of any period in the history of manned flight. More than anyone else, he made Americans an "air-minded" people, prepared them and his subordinate airmen for the rigors of the Second World War, and helped set the stage for the development of an American aircraft industry and flying culture still unmatched in the world. Mitchell's actions are thus simultaneously inspirational, instructive, and troubling—perhaps a fitting legacy for an airman who never shied from a fight and did what he thought necessary to make his ideas about air power a reality.

Robert S. Ehlers Jr.

# FOREWORD

This book is dedicated to those officers and men of the air service who have given up their lives in the development of our national air power.

Few people outside of the air fraternity itself know or understand the dangers that these men face, the lives that they lead and how they actually act when in the air, how they find their way across the continent with unerring exactness—over mountains, forests, rivers and deserts; what they actually do in improving the science and art of flying and how they feel when engaged in combat with enemy aircraft. No one can explain these things except the airmen themselves. The number of these who have had experience and who are capable of expressing themselves, is rapidly growing fewer. Every opportunity should be taken by those that remain to enlighten their fellows on this subject.

The interest in the development of our national air power was manifested by the people of the United States during the past winter, and this interest is growing. The history of the development of air power has been very similar in all countries—it has had to struggle hard to get on its feet. Air power has brought with it a new doctrine of war which has caused a complete rearrangement of the existing systems of national defense, and a new doctrine of peace which eventually will change the relations of nations with each other due to the universal application and rapidity of aerial transport.

This little book has been thrown together hastily. It is compiled from evidence that has been given before the Congress of the United States, articles that have appeared in the public journals and from personal experiences. Its value lies in the ideas and theories that are advanced which it is necessary for our people to consider very seriously in the development of our whole national system. The great countries of Europe have already acted along the lines indicated in this book. We are still backward.

The book is intended to serve several purposes. First, that of putting down in words what the air men think about the organization of an air force and what our national defense should be. Next, to give to the people in general a book which will set before them facts about aeronautical development. And third, a book to which our people in the services, in the executive departments and in Congress can refer for data on aviation which is modern and which is the result of actual experience. So many erroneous doctrines have been enunciated about aviation by the older services that see in the development of air power the curtailment of their ancient prerogatives, privileges and authority, that we consider it time to challenge these proceedings and to make our own views known.

Aeronautics is such a new and rapidly developing science in the world that those concerned in it have not the age, rank or authority which, in the eyes of the older services, entitles them to speak. Most of the data that Congress gets on the subject of aviation comes from officers or agents who are not actual aeronautical officers and who have not come up through the mill of aeronautical experience, both in war and in peace. The airmen have gained their knowledge by actual experience, not by being members of an old well-established service that has gone on in the same rut of existence for decades.

As transportation is the essence of civilization, aviation furnishes the quickest and most expeditious means of communication that the world has ever known. Heretofore, we have been confined to either the earth or the water as the medium of transportation. Now, we can utilize the air which covers both the earth and the water, and the north and south poles, as the medium through which to travel.

With us air people, the future of our nation is indissolubly bound up in the development of air power. Not only will it insure peace and contentment throughout the nation because, in case of national emergency, air power, properly developed, can hold off any hostile air force which may seek to fly over and attack our country, but it can also hold off any hostile shipping which seeks to cross the oceans and menace our shores. At the same time, our national air power can be used in time of peace for some useful purpose. In this it differs very greatly from the old standing armies and navies which, in time of peace, have to be kept up, trained and administered for war only and are therefore a source of expenditure from which little return is forthcoming until an emergency arises.

The time has come when aviation must be developed for aviation's sake and not as an auxiliary to other existing branches. Unless the progressive elements that enter into our makeup are availed of, we will fall behind in the world's development.

Air power has rudely upset the traditions of the older services. It has been with the greatest difficulty that this new and dominating element has gone forward in the way it has. In the future, no nation can call itself great unless its air power is properly organized and provided for, because air power, both from a military and an economic standpoint, will not only dominate the land but the sea as well. Air power in the future will be a determining factor in international competitions, both military and civil. American characteristics and temperament are particularly suitable to its development.

W. M.

In the United States our people have been slow to realize the changed conditions. Isolated as we have been from possible enemies, the people could see little chance for aggression by others. Separated as we are from Europe by the Atlantic, and from Asia by the Pacific which form most certain and tremendously strong defensive barriers, we seemed to be protected by the design of the Almighty. The coming of aircraft has greatly modified this isolation on account of the great range and speed which these agents of communications are developing. Air ships—that is, lighter-than-air crafts, also called dirigibles—have been designed which will go round the world on one charge of fuel. Airplanes have actually flown for 2,500 miles without landing. This fact, added to the development of chemical warfare and the proof that submarines can cover any part of the seas, have diminished the importance of surface seacraft. The vulnerability of the whole country to aircraft as distinguished from the old conditions that obtained when the frontiers or the coast had to be penetrated before an invasion of the country could be made, has greatly interested the people of the nation.

There never has been any lack of interest by the people in aviation in the United States. What has hindered its development has been the extreme conservatism of the executive departments of the government. The pressure of the people on Congress has resulted in the beginning of decisive congressional action.

For the first time in our history, a committee of Congress, during the winter 1924-1925, conducted hearings on every phase of our national defense as affected by air power, while another committee conducted hearings on a specific provision for the creation of a United Air Force, which could be discussed paragraph by paragraph. The evidence taken showed conclusively that the advent of air power has completely changed all former systems of national defense. Air power not only has decisive military advantages, but most of it can be used in time of peace for some useful purpose.

The hearings before these committees were reported in the public press throughout the country, and, for the first time, it was brought home to our citizens what air power means. The frontiers in the old sense —the coast lines or borders—are no longer applicable to the air because aircraft can fly anywhere that there is air. Interior cities are now as subject to attack as those along the coast. Nothing can stop the attack of aircraft except other aircraft.

The evidence shows plainly that the United States has adopted no modern plan of organization for meeting the general world movement in the organization of its air power. It still adheres to the methods and systems of many years ago. This has resulted in a very much retarded development of our aeronautical resources entirely out of proportion with the aeronautical capabilities of our country. We lead the world in undeveloped aeronautical material, our men make the best flyers and mechanics, our factories are capable of turning out the best airplanes, and we have all the raw materials that are necessary.

From a military standpoint, no specific mission is assigned to our national aeronautics. It is regarded as an auxiliary to the Army and Navy. Actually there is no air force in the United States. The system of creating one is so complicated and so difficult to put in motion that an air force could only be brought to a state of efficiency after years of trial, hundreds of mistakes, and the wasting of many lives and millions of dollars in money. In other words, relatively we are little better off than we were at the beginning of the World War.

Rapidity of modern means of communication, the sureness of various means of transportation, and the accessibility of all parts of the world to aircraft, which have been developed in an incredibly short space of time, make it absolutely necessary that we organize to meet modern conditions. Our various means of national defense must be accurately coördinated because the next contest will increase the swiftness with which decisions are reached and the nation that hangs its destiny on a false preparation will find itself hopelessly outclassed from the beginning.

Neither armies nor navies can exist unless the air is controlled over them. Air forces, on the other hand, are the only independent fighting units of the day, because neither armies nor navies can ascend and fight twenty thousand feet above the earth's surface.

The missions of armies and navies are very greatly changed from what they were. No longer will the tedious and expensive processes of wearing down the enemy's land forces by continuous attacks be resorted to. The air forces will strike immediately at the enemy's manufacturing and food centers, railways, bridges, canals and harbors. The saving of lives, man power and expenditures will be tremendous to the winning side. The losing side will have to accept without question the dominating conditions of its adversary, as he will stop entirely the manufacture of aircraft by the vanquished.

Surface navies have entirely lost their mission of defending a coast because aircraft can destroy or sink any seacraft coming within their radius of operation. In fact, aircraft today are the only effective means of coast protection. Consequently, navies have been pushed out on the high seas. The menace of submarines from below and aircraft from above constitutes such a condition that the surface ship as an element of war is disappearing. Today, the principal weapon in the sea is the submarine with its mine layers, gun fighters and torpedo craft.

In the future, campaigns across the seas will be carried on from land base to land base under the protection of aircraft. Expeditions across the sea such as occurred in the World War will be an impossibility. Water spaces between land bases in the northern hemisphere are very short. The space from America to Asia is only fifty-two miles across the Bering Straits and across the Atlantic it is scarcely more than four hundred.

Air power can hold and organize small islands in a manner which has been entirely impossible heretofore. These can be supplied by other aircraft, or by submarines, with everything that is necessary.

Should it be required to use surface ships, merchantmen may be taken and equipped with flying-off decks for use as airplane transports whenever the necessity arises. Consequently, the power of navies as a keystone in the arch of national defense has been relegated to a secondary position.

Each nation in the world is fully or partially recognizing these principles and is organizing its national defense system accordingly. What is necessary in this country is that the people find out the exact conditions concerning air power and the exact truth about what it can accomplish in time of peace as well as in time of war.

Every man who has had experience in the air is able to aid in this work. It is the biggest constructive program that we have before us. A reasonable solution of this problem clearly indicates that we should have a single Department of National Defense and under it a Department of Aeronautics, Department of the Army, and Department of the Navy. The views of the air must be heard in the national councils on an equal basis with those of the Army and Navy.

The mission of a Department of Aeronautics should be to provide for the complete aeronautical defense and aeronautical development of the country; that of the Army, for the protection of the land areas; that of the Navy, for naval operations out on the high seas and not along the coasts nor on the land.

A United Air Force would provide an aeronautical

striking force designed to obtain control of the air and demolish whatever hostile land or water targets might be necessary, according to the military situation.

The personnel situation is very serious in all the Air Services. The air-going people actually form a separate class. They are more different from landsmen than are landsmen from seamen. At the present time, the air-going people in the national services are not accorded the position nor the rank to which the importance of their duties entitles them. Many officers in the Air Service with the rank of major, captain and even lieutenant, are charged with responsibilities even greater than those of generals in the Army or Admirals in the Navy.

Their position on the promotion list is hopeless. Some of our lieutenants can never rise above the rank of major or even captain. They see no future before them and consequently are not in the state of mind in which officers in so rapidly developing a service should be. Without satisfied, energetic, capable personnel, no air service can be developed.

Changes in military systems come about only through the pressure of public opinion or disaster in war. The Army and Navy have regularly organized publicity bureaus which can disseminate information about these services, but there is no medium through which essentially aeronautical information can be disseminated. The result is that the public and Congress are slow to get all the aeronautical facts.

The evidence before both Congressional Committees plainly showed that:

1. There should be a Department of Aeronautics charged with the complete aeronautical defense and the aeronautical development of the country.
2. There should be an aeronautical personnel entirely apart from the Army and Navy.
3. There should be a Department of National Defense with sub-heads for the Air, Army and Navy.

It remains for Congress to translate these principles into law.

CONTENTS

# WINGED DEFENSE

# WINGED DEFENSE

## I

## THE AERONAUTICAL ERA

THE world stands on the threshold of the "aeronautical era." During this epoch the destinies of all people will be controlled through the air.

Our ancestors passed through the "continental era" when they consolidated their power on land and developed their means of communication and intercourse over the land or close to it on the seacoast. Then came the "era of the great navigators," and the competition for the great sea lanes of power, commerce, and communication, which were hitched up and harnessed to the land powers created in the continental era. Now the competition will be for the possession of the unhampered right to traverse and control the most vast, the most important, and the farthest reaching element of the earth, the air, the atmosphere that surrounds us all, that we breathe, live by, and which permeates everything.

Air power has come to stay. But what, it may be asked, is air power? Air power is the ability to do something in or through the air, and, as the air covers the whole world, aircraft are able to go anywhere on the planet. They are not dependent on the water as a means of sustentation, nor on the land, to keep them up. Mountains, deserts, oceans, rivers, and forests, offer no obstacles. In a trice, aircraft have set aside all ideas of frontiers. The whole country now becomes the frontier and, in case of war, one place is just as exposed to attack as another place.

Aircraft move hundreds of miles in an incredibly short space of time, so that even if they are reported as coming into a country, across its frontiers, there is no telling where they are going to go to strike. Wherever an object can be seen from the air, aircraft are able to hit it with their guns, bombs, and other weapons. Cities and towns, railway lines and canals cannot be hidden. Not only is this the case on land, it is even more the case on the water, because on the water no object can be concealed unless it dives beneath the surface. Surface seacraft cannot hide, there are no forests, mountains, nor valleys to conceal them. They must stand boldly out on the top of the water.

Aircraft possess the most powerful weapons ever devised by man. They carry not only guns and cannon but heavy missiles that utilize the force of gravity for their propulsion and which can cause more destruction than any other weapon. One of these great bombs hitting a battleship will completely destroy it. Consider what this means to the future systems of national defense. As battleships are relatively difficult to destroy, imagine how much easier it is to sink all other vessels and merchant craft. Aerial siege may be laid against a country so as to prevent any communications with it, ingress or egress, on the surface of the water or even along railways or roads. In case of an insular power which is entirely dependent on its sea lanes of commerce for existence, an air siege of this kind would starve it into submission in a short time.

On the other hand, an attempt to transport large bodies of troops, munitions, and supplies across a great stretch of ocean, by seacraft, as was done during the World War from the United States to Europe, would be an impossibility. At that time aircraft were only able to go a hundred miles before replenishing their fuel; now they can go a thousand miles and carry weapons which were hardly dreamed of in the World War. For attacking cities that are producing great quantities of war munitions that are necessary for the maintenance of an enemy army and country in case of war, the air force offers an entirely new method of subduing them. Heretofore, to reach the heart of a country and gain victory in war, the land armies always had to be defeated in the field and a long process of successive military advances made against it. Broken railroad lines, blown up bridges, and destroyed roads, necessitated months of hardships, the loss of thousands of lives, and untold wealth to accomplish. Now an attack from an air force using explosive bombs and gas may cause the complete evacuation of and cessation of industry in these places. This would deprive armies, air forces, and navies even, of their means of maintenance. More than that, aerial torpedoes which are really airplanes kept on their course by gyroscopic instruments and wireless telegraphy, with no pilots on board, can be directed for over a hundred miles in a sufficiently accurate way to hit great cities. So that in future the mere threat of bombing a town by an air force will cause it to be evacuated, and all work in munitions and supply factories to be stopped.

A new set of rules for the conduct of war will have to be devised and a whole new set of ideas of strategy learned by those charged with the conduct of war. No longer is the making of war gauged merely by land and naval forces. Both of these old, well understood factors in conducting war are affected by air power which operates over both of them. Already, we have an entirely new class of people that we may call "the air-going people" as distinguished from the "land-going people" and the "sea-going people." The air-going people have a spirit, language, and customs of their own. These are just as different from those on the ground as those of seamen are from those of land men. In fact, they are much more so because our sea-going and land-going communities have been with us from the inception of time and everybody knows something about them, whereas the air-going people form such a new class that only those engaged in its actual development and the younger generation appreciate what it means.

The airmen fly over the country in all directions constantly, winter and summer they go, as well as by night and by day. The average dweller on the earth never knows that above him aircraft in the United States are speeding between the Atlantic and Pacific and from the northern frontier to the southern frontier, on regular scheduled trips. The pilots of these planes, from vantage points on high, see more of the country, know more about it, and appreciate more what the country means to them than any other class of persons.

Take, for instance, a trip from the east coast out to the Middle States, accomplished in four or five hours. One starts in the morning from the Atlantic. Looking out across it for miles along the coast, the shipping coming from Europe can be plainly seen entering the harbors. Back from the coast itself stretch the industrial cities with their great factories, pushing out to the West; numberless steel lines of railways searching for the gaps in the mountains to take them through to the Middle States; the strip of cities with their heavy populations is passed, then the small farms straggling into the Alleghany Mountains; with the white roads growing fewer and fewer as the Blue Ridge Mountains are approached. Once into the Alleghanies, the utter lack of development makes itself evident at once; as far as the eye can reach there is scarcely a habitation, a road, or a clearing. The inhabitants, deprived of the means of communication, are probably our least educated citizens, although largely the purest-blooded Americans in the country. Across the Alleghanies we reach the rich lands of the Middle States. The great farms seem to crowd themselves against each other in order to produce the largest crops.

The country is traversed by well-made roads, railroads, electric power, telegraph, and telephone lines. Bright, clean cities are dotted with splendid schools, fine public works, parks, and hospitals. The development of the animal industry is tremendous; cattle, pigs and sheep are in abundance. While interspersed in this great agricultural country, we still find great cities with high chimneys belching black smoke, indicating the presence of great industries.

A few hours more and the airplane traverses the whole country to the Pacific Coast. Certainly no other class of men appreciate their country or know so much about it as the "air-going fraternity."

The absorbing interest in this new development is so great that the youth of the country everywhere is being inspired to make this their specialty. Bold spirits that before wanted to "go down to the sea in ships," now want to go "up in the air in planes."

The air force has ceased to remain a mere auxiliary service for the purpose of assisting an army or navy in the execution of its task. The air force rises into the air in great masses of airplanes. Future contests will see hundreds of them in one formation. They fight in line, they have their own weapons and their own way of using them, special means of communications, signalling, and of attacking.

Armies on the ground or ships on the water have always fought on one surface because they could not get off it. The air force fights in three dimensions—on the level, from up above, and from underneath. Every air attack on other aircraft is based on the theory of surrounding the enemy in the middle of a sphere with all our own airplanes around the whole periphery shooting at it. If we attack a city or locality, we send airplanes over it at various altitudes from two or three hundred feet up to thirty thousand, all attacking at once so that if any means of defense were devised which could hit airplanes or cause them to be destroyed from the ground, the efforts would be completely nullified, because they could neither see, hear, nor feel all of them. No missile-throwing weapons or any other devices have yet been created or thought of which can actually stop an air attack, so that the only defense against aircraft are other aircraft which will contest the supremacy of the air by air battles. Great contests for control of the air will be the rule in the future. Once supremacy of the air has been established, airplanes can fly over a hostile country at will.

How can a hostile air force be forced to fight, it may be asked, if they do not desire to leave the ground? The air strategist answers: "By finding a location of such importance to the enemy that he must defend it against a bombardment attack by airplanes."

Such a place as New York, for instance, would have to be defended if attacked by hostile bombers, and, as no anti-aircraft guns or other efforts, from the ground alone, would be of any particular avail, aircraft would have to be concentrated for its defense and a succession of great air battles would result. Putting an opponent on the defensive in the air is much more valuable comparatively than putting him on the defensive on the ground. Armies may dig trenches, live in them, or sit around in them waiting for an enemy to attack them. This cannot be done in the air for airplanes have to return to the ground periodically for refueling. If they are not in the air when the hostile air force appears, they will have no effect on it, because they cannot arise to a great altitude and catch it. Consequently, not more than about one-third of an air force can be kept constantly in the air, so that in the future, the country that is ready with its air force and jumps on its opponent at once will bring about a speedy and lasting victory. Once an air force has been destroyed it is almost impossible to build it up after hostilities commence, because the places capable of building aircraft will be bombed and the big air stations that train pilots and flyers will be destroyed. Even if the country on the defensive is able to create small parcels of aviation, they will be destroyed in detail, one after the other, by the victorious air force which not only has control of the air but is protecting its own interior cities that manufacture and turn out their equipment, airplanes, and supplies.

From an aeronautical standpoint, there are three different classes of countries: First, those which are composed of islands subject to air attack from the coast of a continent. In this case the insular country must completely dominate the air if it wishes to use an army against its neighbors so as to be able to transport and land it on the shores of the Continent. If its opponents on the Continent control the air, they can cut off all the insular country's supplies that come over the seas, they can bomb its ports and its interior cities, and, with their air force alone, bring the war to a close.

The second class of country is the one that has a land frontier directly facing and joining its opponent and which is partially self-sustaining and partially dependent on food and supplies from outside, either by railways, by sea, or by air. In this case, there is a possibility that armies might come into hostile contact if the air forces did not act quickly enough. Even then if the air forces of one or the other were ready at the start of the war, all the important cities would be laid waste, the railroads and bridges destroyed, roadways constantly bombed and torn up so as to prevent automobile transportation, and all seaports demolished. Again the air force might bring victory unaided to the side which was able to control the air.

The third class of country is one which is entirely self-sustaining but is out of the ordinary aircraft range. The United States comes under this category. No armed force of an European or Asiatic nation can come against the United States except through the air or over the water. An efficient air force in this instance would be able to protect the country from invasion and would insure its independence but would not be able to subject a hostile country to invasion, or to defeat it without leaving the country itself.

Consequently, an entirely new method of conducting war at a distance will come into being. We have seen that a superior air power will dominate all sea areas when they act from land bases and that no seacraft, whether carrying aircraft or not, is able to contest their aerial supremacy.

Strings of island bases will be seized by the strong powers as strategic points so that their aircraft may fly successively from one to the other and as aircraft themselves can hold these islands against seacraft, comparatively small detachments of troops on the ground will be required for their maintenance. An island, instead of being easily starved out, taken or destroyed by navies as was the case in the past, becomes tremendously strong because it cannot be gotten at by any land forces and, while supremacy of the air is maintained, cannot be taken by sea forces.

In the northern hemisphere there is no stretch of water greater than the present cruising range of airplanes that has to be crossed in going from America to Europe or from America to Asia. The farther north we go the narrower the intervals of water between the continents. The Behring Straits are only fifty-two miles wide, while in their center are two islands that make the widest stretch twenty-one miles, scarcely more distance than across the English Channel. The greatest straight line distance over the narrowest stretch of water between America and Europe is about four hundred miles, or four hours flight.

Cold is no impediment to the action of aircraft. In fact, the colder the weather, the clearer the sky and the better the flying conditions. The sun's rays are what make most of the trouble for the aviator. In the first place, they cause heat, which makes the air hold more water. When the air cools it causes fogs, clouds and haze, because the moisture congeals as the air can no longer hold it. The heat from the sun causes ascending currents of air and the air around rushes in to take the place of the ascending currents. This makes storms of all kinds—causes what we used to call holes in the air, which are merely up and down currents, and introduces much the same difficulties that storms at sea cause to ships.

Light also interferes with our radio or wireless telegraph and telephone communication. Radio waves are really elongated light undulations, and whenever there is light in the air, we hear some of the overtones and undertones from it. That is why the best time for radio telegraphy is at two or three o'clock in the morning when all the light has gone out of the air and before more light has come. That is also the best time for flying because on account of the coldness of the night, the moisture has been deposited on the earth, the absence of light and heat has ceased to make up and down currents in the air and there are no heavy winds. This is the reason why all migratory birds, knowing this, fly at night in their migrations from north to south. It really is easier for the airman to fly at night instead of in the daytime, and in the future much of our traffic, especially for all heavy planes, will be conducted at night. Ice and snow cause the little holes, furrows, and ridges in the ground to be filled with a soft substance that makes natural airdromes everywhere and the sheets of water are turned into ice which can be utilized for landing.

Our aerial routes between the continents will not

follow the old land and water ways parallel to the equator which have been used heretofore, because our old means of transportation used to be confined to land and water in warm parts of the earth. The new routes will follow the meridians, straight over the top of the earth, which cut off hundreds of miles, save weeks of time, untold effort, dangers and expense.

What will this new element in warfare result in? Unquestionably, the amelioration and bettering of conditions in war because it will bring quick and lasting results. It will require much less expense as compared with that of the great naval and land armies which have heretofore been the rule and it will cause a whole people to take an increasing interest as to whether a country shall go to war or not, because they are all exposed to attack by aircraft, no matter if they live in the remotest interior of the country.

Now, much of a country's population thinks because it does not live near a seacoast or a land frontier, that its homes will be safe from attack and destruction. The worst that can happen to them, in case of defeat in war, would be higher taxes to pay, and war debts, because navies cannot reach them and armies only with the greatest difficulty. Let us look back and see what warfare used to be and how it evolved.

Primitive man fought his neighbor with his teeth, his hands, and feet. His adversary was killed in the struggle. Great individual fighters developed who were stronger than their fellows. Next, the man obtained a club with which to hold his enemy off at a distance. Then came the thrown missile, such as a stone. Then, getting others to assist him, which in time resulted in what we call "armies" today. Good steel weapons were invented. Great armies were created using steel. This brought them body to body in their contests. Those who were vanquished, were entirely destroyed; their cities looted and burned, and the whole country laid waste. At that time an entire country went to war. The men fought the enemy while the women and children supplied their wants, manufactured their clothing, and accompanied them on their marches. Gradually, the method of working steel became so excellent that armor could be made which would resist the attack of all known weapons, but, as armor cost a great deal and was hard to get, it developed into a few armed men doing all the fighting for their people. Instead of the armies being universal service institutions in which every man had to take part, as was formerly the case, only a few did the fighting while the others worked at their civil occupations. The advent of gunpowder changed all of this. The knight could no longer resist the peasant armed with a musket and, gradually, all the fighting nations were organized again so that all of their man power could be called to the colors or into the workshops when war was declared. This is the condition that exists today in all countries. The armies themselves, their operations, their strategy, and even their tactics are little different from what they were in the days of the Romans.

As weapons have been improved and made more terrible, such as the long range cannon, the machine gun, and toxic gas, just so much have the total casualties and losses been reduced, because the enemy and those engaged in combat are held further apart. Victories are sharp and decisive because it can be seen what the results will be, long ahead of time, and the defeated side can get away with its men as they are far off from their opponents and not body to body as they were in the days of the Romans.

The Great War in Europe, barely finished at the present time, was not as severe a contest for the fighters as our own Civil War in America was sixty years ago. The casualties at that time were much heavier in proportion to the numbers engaged. The proportion of the population serving under the colors of the armies was also much greater, and the utter destruction of the vanquished states in the South was ten times worse than anything that happened in Europe. This was because in 1914-1918 weapons of greater range were used—the machine gun gave greater defensive strength, and the men fighting were held farther apart.

As air power can hit at a distance, after it controls the air and vanquishes the opposing air power, it will be able to fly anywhere over the hostile country. The menace will be so great that either a state will hesitate to go to war, or, having engaged in war, will make the contest much sharper, more decisive, and more quickly finished. This will result in a diminished loss of life and treasure and will thus be a distinct benefit to civilization. Air forces will attack centers of production of all kinds, means of transportation, agricultural areas, ports and shipping; not so much the people themselves. They will destroy the means of making war because now we cannot cut a limb from a tree, pick a stone from a hill and make it our principal weapon. Today to make war we must have great metal and chemical factories that have to stay in one place, take months to build, and, if destroyed, cannot be replaced in the usual length of a modern war.

Navies, it is interesting to note, came into being as organized units as parts of armies to be used merely as the vehicle of transportation of the soldier so that he could get at close grips with the enemy and determine the possession of the sea areas. As long as the boats were propelled by oars and could go where the army officers told them, with certainty, they remained under the control of the army. Only within the last couple of centuries have the navies become independent of the armies. This came when navies used sails for propulsion and they could not tell with certainty whether they could go where they were ordered or not. Now that steam and the internal combustion engines have come for propelling seacraft, both surface and subsurface, and their power in war has been so tremendously curtailed by the advent of aircraft, it is probable that they will again revert to being an auxiliary of armies and air forces.

A comparatively small part of a population ever serves in navies, and, compared to armies, they, alone, practically never bring a war to an end. They have acted as auxiliaries to an army in clearing the sea of enemy ships, so as to be able to transport an army or to assist it in getting close to its enemy.

In considering the relations between armies, navies,

and air forces, we may say that the armies have reached an epoch of arrested development in which the controlling factors, as they have always been, are a man's physical strength, his power to march, and his power to see. The use of his weapons entirely depends on these attributes. Their augmentation by mechanical transportation and raised platforms for observation does not alter this general condition. Of course, everything begins and ends on the ground. A person cannot permanently live out on the sea nor can a person live up in the air, so that any decision in war is based on what takes place ultimately on the ground.

The rôle of armies and their way of making war will remain much the same in the future as it has in the past, if air power does not entirely prevent them from operating.

Navies, however, are able to control only the areas of water outside of the cruising radius of aircraft. These areas are constantly diminishing with the increasing flying powers of aircraft. It will be impossible for them to bombard or blockade a coast as they used to, or ascend the rivers, bays, or estuaries of a country adequately provided with air power.

The surface ship, as a means of making war, will gradually disappear, to be replaced by submarines that will act as transports for air forces and destroyers of commerce.

The advent of air power holds out the probability of decreasing the effort and expense required for naval armaments, not only in the craft themselves, but in the great bases, dry docks, and industrial organization that are necessary to maintain them. Differing from land armies, which are in a stage of arrested development, navies are in a period of decline and change. The air force is the great developing power in the world today. It offers not only the hope of increased security at home, but, also, on account of its speed of locomotion, of the greatest civilizing element in the future, because the essence of civilization is rapid transportation. It is probable that future wars again will be conducted by a special class, the air force, as it was by the armored knights in the Middle Ages. Again the whole population will not have to be called out in the event of a national emergency, but only enough of it to man the machines that are the most potent in national defense.

Each year the leading countries of the world are recognizing the value of air power more and more. All of the great nations, except the United States, have adopted a definite air doctrine as distinguished from their sea doctrine and their land doctrine. To develop anything, the underlying thought and reason must govern and then the organization must be built up to meet it. The doctrine of aviation of all of these great countries is that they must have sufficient air power to protect themselves in case they are threatened with war. Each one solves the matter in a way particularly adapted to its own needs.

All of them started out by having the aviation distributed under many different heads—the army having its part, the navy having its part, the civil and commercial aviation their parts, airplane constructors having another part, the weather or meteorological service and wireless communications still another. All of these services considered aviation as auxiliary or subsidiary to some other activity whose principal application was not aviation. Just as the navy always thinks first of the battleships and makes aviation secondary to that, the army thinks of the infantry and also makes aviation a secondary matter.

The armed services of a nation are the most conservative elements in its whole makeup. To begin with, they antedate the governments themselves, because all governments have been brought into being by great popular upheavals which have found expression in military forces. The traditions among all the armed services are much older than any government, more conservative than any department of government, and more sure to build on a foundation that they are certain of rather than to take any chances of making a mistake. As they have changed so little in their methods and ways of conducting war for so many centuries, they always look back to find a precedent for everything that is done.

Hindenburg looked back to Hannibal's battle of Cannæ, and made his dispositions to fight the Russians at Tannenberg. Napoleon studied the campaigns of Alexander the Great and Genghis Khan, the Mongol. The navies drew their inspiration from the battle of Actium in the time of the Romans and the sea fight of Trafalgar.

In the development of air power, one has to look ahead and not backward and figure out what is going to happen, not too much what has happened. That is why the older services have been psychologically unfit to develop this new arm to the fullest extent practicable with the methods and means at hand.

The trend in all nations has been to centralize their aeronautical efforts with a view of developing aviation for aviation's sake first; next, to cut out all the duplication and expense incident to having several agencies do the same thing.

Great Britain leads the world in this conception of air power. She now has an Air Ministry which is co-equal with the Army and Navy. Her air force is designated by law as "the first line of defense" of the United Kingdom. The country is completely organized into aeronautical defense areas with the pursuit and bombardment aviation all under one command, so that the maximum power may be brought to bear anywhere desired, and not have it split up between the army and navy as it used to be. In addition to this there are home defense air forces assigned for the permanent defense of London and other important cities. In case of war these would never leave their posts. Radiating out from these aviation centers are listening and operation posts all along their coasts and even out at sea, so that any hostile aircraft approaching will be promptly reported. The air force personnel of air officers are in part on permanent duty with the air force, and might be called "regulars," and the other part are in the "Reserve." The reserve officers fly for a short time each week or month and turn out for a period of a couple of weeks each year with their organizations.

The British air forces have their aeronautical academy, corresponding somewhat to our West Point for the army in America. They also have their staff colleges and all the machinery necessary for a great army of national defense. Its importance is growing every day.

From an industrial side, they foster and encourage the engineers for aircraft and the factories that make them. The money that is appropriated for building airplanes is used so as to keep up and maintain the best factories for the production of aircraft. Civil aviation receives great consideration. It is not yet as economical a means of transportation as railways or steamboats, so it has to be helped by the Government. Civil aviation uses the same kind of landing fields and airdromes as military air forces. The great majority of these are maintained in time of peace by civil aviation, and are always ready in case of war. Large subsidies and bounties are paid to the air transportation companies so that they may be able to operate and gain a small profit. In this way the Government keeps many pilots, machines, airplanes, and mechanics doing air work in time of peace at only half the cost of what it would be if they were engaged entirely on Government work.

Now it is reported that the organization of Great Britain's whole military force has gone so far as to make an air officer responsible for the whole defense of the British Isles. In case of a war in the future, this air officer will have under his orders not only the air force, but also the army and navy for the protection of the Islands. This may be extended to the whole Empire at a later date.

An air officer was selected because his training gives him an insight into the land and sea operations, which no other service can possess. His means of reconnaissance and gaining information of an enemy for hundreds of miles away from his frontiers is greater than any means possessed by either an army or navy. His air forces move many times as fast as any ground or water service, consequently he is in a better position to know where an enemy will hit and what measures should be taken to protect his country and combine everything in the national defense—air, land and water. This also makes it possible for a state to hold one man responsible for the conduct of the national defense and not have the duty divided between entirely separate commands handling air, land and water forces.

In Mesopotamia, Irak as it is called, the air force handles the military occupation of the country in a manner similar to that in which armies have in the past. The result of this occupation has been very satisfactory. The airplanes fly over the country at will, are able to put down uprisings quickly, transport troops to places where they are needed on the ground, and to cover much more country with less effort than is possible by any other means. In this area, all the detachments of the army are made auxiliary to the air force, and are under the air force commander. The great nations of Europe and Asia are now approximating this organization more and more, as it becomes increasingly evident that air power, to be given its maximum chance, must be developed as a main arm instead of as an auxiliary.

Not every nation is capable of developing an efficient air force. To create one, two things are necessary. First, a strong national morale, a patriotism and love of country which will impel its pilots to withstand tremendously high losses in case of war. Only a few nations have this power. China, for instance, is organized on the basis of family, commercial relations, and a biological supremacy, rather than on a pedestal of national defense by armed forces to keep foreign nations from disturbing her. She cannot create an efficient military aviation at this time because there is no central Government upholding the basic principles or maintaining the ideals which the intelligent people are willing to give up their lives and their all for. On the other hand, the American aviation at the battle of Château Thierry, with seventy-five per cent of its strength killed, wounded, and missing in little over two weeks, kept right on fighting with as great a morale as if these losses had not occurred. Suitable pilots can be drawn only from certain classes, such as the young men who go to our colleges and not only are proficient in their studies, but in athletics such as football, baseball, tennis, polo, and other equestrian exercises which make the body and mind act together quickly. The United States has the greatest reservoir of this kind of personnel of any nation in the world.

The second important element in the creation of an air force is the industrial condition of a country and its supply of raw materials that go into the creation of aeronautical equipment, engines, and airplanes. Seventy-eight different trades are represented in the building of a single plane. From the time it is devised until the time it is turned out in production, it takes as long as it does to build a battleship. Everything in the airplane revolves around the engine. Again we find very few countries capable of manufacturing suitable aeronautical engines. Think what is involved in this —the mining of all metals, their conversion into the toughest and lightest alloys that are known to science, then the designing, building and testing of these engines that weigh scarcely over a pound to the horse power and that are capable of pulling an airplane through the air once they are in flight, with hardly the use of any wings. Again using China as an example we find that no aeronautical engines are made in that country, nor any internal combustion engines for that matter, as her industries have never been organized along those lines. The United States, on the other hand, has the greatest motor industry on the earth, in the form of automobile manufacturers. These are conversant with all phases of the internal combustion, gasoline engine. For this reason we lead the world in the excellence of our aeronautical engines at the present time. At the same time the United States has within its own borders all the raw materials, fuels, and expert laborers that are necessary in the production of aeronautical equipment.

If a nation ambitious for universal conquest gets off to a "flying start" in a war of the future, it may be able to control the whole world more easily than a nation has controlled a continent in the past. The advent of air power has made every country and the world smaller. We do not measure distances by the unit of miles, but by the unit of hours. Communications all over the world today are instantaneous, either by the submarine telegraph, by the land line, or by radio telegraphy. Airplanes can be talked to while in flight anywhere. The airship or Zeppelin can cross any ocean. Should a nation, therefore, attain complete control of the air, it could more nearly master the earth than has ever been the case in the past.

Just as power can be exerted through the air, so can good be done, because there is no place on the earth's surface that air power cannot reach and carry with it the elements of civilization and good that comes from rapid communications.

# II

# LEADERSHIP IN AERONAUTICS GOES TO THE UNITED STATES

"THERE is no royal road to learning." This old saying holds more true in aviation than in any other calling.

In the old and well established branches of learning there is something to go on that has been developed before, that one can model on and study. In aviation, particularly in its application and use, there is almost nothing to go on. The air man has to "learn" himself, for the most part. Every new development, no matter what it is, requires the greatest preparation beforehand to insure its success and with us in aviation it has been essential that we be successful or we might not be allowed to carry on our experiments to the point where they would be of the utmost value to the country. We have constantly before us the example of Langley, who was on the verge of flying his heavier-than-air machine when the ridicule of some caused Congress to withhold funds, and to stop one of the most important single accomplishments that has occurred in the world's history. Fortunately Congress has been pretty good to aviation. Whenever it could see the why and the wherefore, and understood that was actually needed in aviation, it has provided the necessary funds. Furthermore, Congress has a good deal of vision, that is, seeing what might happen and being willing to gamble a little on the result.

A great sum of money was appropriated for aviation during the war and Congress expected immediate results, which, of course, were not forthcoming. At the end of the war it was beginning to be understood that money could not buy knowledge of aviation; that this was a matter of development that required time; that the fault had been not so much in the spending of the appropriation, but in not laying a sound foundation for the spending of the money before the European war started.

Practically nothing had been done by the Government to start a real air service. There were only about fourteen actual flyers when the war in Europe commenced. The prodigious effort put forth by the United States to create an aviation during the war had barely time to show results in the year and nine months of the contest. There was created, though, an actual flying personnel; fifteen thousand of our citizens had received instruction in flying. These men were the finest pilot material in the world. Great numbers of factories were making aircraft, and no matter what may have been the reasons for the kinds of aircraft that they were building, they were making the types given them to construct better and more cheaply than any in Europe. Only a few of the flying men and a relatively small part of the material had gotten to Europe when the war ceased. We were still using foreign equipment because there had not been sufficient time to create our own.

It had been conclusively shown that aviation was a dominant element in the making of war even in the comparatively small way in which it was used by the armies in Europe. If either one of the opposing forces had been deprived of their aviation and the routes of the air had been perfectly clear to the other side, the side having the aviation would have been victorious within a couple of weeks.

The European War was only the kindergarten of aviation. It had machines that were just invented, the possibilities of their use were just beginning to be understood by the aviators themselves, while others looked on them as strange creations that were defying all known laws of science, of custom and of war.

During the year 1918, the American aviation received its baptism of fire in a terrific manner. Hurled in the midst of the Château Thierry struggle, with the Allies shot out of the air, we had to evolve our own system and salvation as best we could. Untired by three long years of war, our men thought out many new ways of applying and handling air forces, so that when the armistice came we had a fighting staff entirely conversant with the last minute ways of making air power felt, and pilots that had fought the hardest of air battles, that knew every kink of the air fighting game, and that knew they could defeat in single or combined combat any aviators of the world.

In addition, we had handled the greatest combined force of aviation ever brought under one command during our St. Mihiel and Argonne operations. When the question was agitated of consolidating the command of water, land, and air under one direction on the Western Front for the campaign of 1919, it is probable that all sea forces would have been commanded by the British, all land forces by the French, and all air forces by the Americans. The power of aeronautical accomplishment was passing to the United States and without a doubt it would have had the utmost influence in settling the war had it lasted another year.

During the spring of 1919, our war-trained men came back from Europe. Most of them were discharged, many could see no future under the conditions under which aviation was being handled in this country, and others wished to return to private life. Enough remained in service, however, to lay the foundation of our future aviation. These men, mixed with the aviation officers that remained in the United States, who had been engaged in training pilots and testing equipment to send to Europe, formed the combination that was capable of thinking out what our air organization should be, more than any other class of men. They knew the actual conditions that existed and had sufficient vision to see what aviation could do in the future.

To the air officer, the conditions of a future war would be entirely and wholly different from anything that occurred in Europe. The armies in Europe at the commencement had no aviation. As each side developed its aviation, it maintained equality in the air, more or less, until within three or four months of the Armistice, when we were gaining a tremendous advantage; sufficient advantage, certainly, to determine the outcome of the war in 1919. On the other hand, any campaign in the United States would involve first a defense of our coast against foreign air or sea forces coming from Europe or Asia. In this, the ground army would have very little part and a properly organized air force could protect the country, not only the frontiers and seaports, but the smallest hamlet in the highest mountains, because all are exposed to air attack. If it became necessary to impose our will on an enemy, the campaign would have to be carried to him through the air because an effective air force on his part would prohibit the transportation across the water of armaments that has been possible in the past. Consequently, our development must be based on the grand hypothesis that future contests will depend primarily on the amount of air power that a nation could produce and apply.

Our own mission in aviation, as Air Officers of the United States, was to demonstrate its practicability, dependability, and efficiency.

The elements of air power are very numerous and complicated. To begin with, the personnel: officers, mechanics, designers, manufacturers, engineers, and inspectors, all have to be created especially for aviation work. This requires a long period of time. It must be based on the proper system of training, while the training, itself, must be based on how air power is to be used. The work of an air force depends on the men that fly the planes, not primarily on those that remain on the ground. The avenues along which military, civil and commercial air power is to be developed must be selected. What we call airways must be organized just as roads had to be laid down for automobiles and refueling stations installed; or, as lines for steamships had to be established with ports where fuel, coal, and oil could be obtained. Just so do our airways have to be made for air power. These airways can be used for both commercial and military planes. In the future we will see the merchant aeronautics alongside the military aeronautics, one being a direct assistance to the other, both using the same airways, the same navigating instruments, and the same methods of flying as the other.

In Europe, during the war, there were no airways because the front was so short that a flight of about two and one-half hours would carry an airplane, even a slow one of that time, from the Atlantic Ocean to the Swiss frontier. The United States had to be ready to organize our airways from the Atlantic to the Pacific, a distance of three thousand miles; and from the northern frontier to the southern frontier, a distance of two thousand miles. The airways had to be connected up by radio communications because we had found that the telephone and telegraph were too slow. We had to put in a weather service so that we could tell thirty-six hours ahead of time what the weather would be and then notify the pilots of the planes in sufficient time that they might be forewarned of fogs, storms or hurricanes. We had to show that we could distribute gasoline and oil, mechanics, and spare parts, to the places selected for landing and we had to show that the whole airway could be controlled from one point, so that in case an air force were stationed in the central part of the country it could move with great rapidity to either coast. More than anything else we had to show that it was practicable for airplanes in considerable numbers even to make a trip from one coast to the other. Most people were still of the opinion that planes could only go a short distance, then would have to stop, be repaired and overhauled. They thought they could fly only in clear weather and that storms, rains, and fogs would stop them. The successful demonstration of the movement of an air force would enable us to control the air along our frontiers and coasts without doubt.

But, even if we were successful in flying across the country, it remained to be demonstrated that we could sink battleships, because, in Europe, the war had been won on land. The Allies held the sea to a greater extent than any nation ever has in the past. Aircraft, therefore, had not been used to any appreciable extent against shipping. It had to be demonstrated that we could fly over the water as easily as we could fly over the land, and that we could work at night as well as by day. We had to perfect weapons and sights for hitting objects on the ground and water and methods of flying in all kinds of weather, in all climates, and under all conditions.

To put these things into effect we had almost no precedent to follow. We had to think them out for ourselves and then proceed to put our ideas and theories to the test. Many of the appliances and the equipment had to be made for the first time and a great deal of the equipment we possessed had to be used in a manner for which it was not designed because all of the aeronautical equipment had been made for service on the western front in Europe. Fortunately, we had the Liberty engine which had been perfected and had become the most reliable aeronautical engine in the world. We had a great accumulation of aviation material: guns, bombs, and instruments, which had never been sent to Europe and which, if not used within a short time, would deteriorate so that it would be useless.

The greatest of all of our assets, however, were the wonderful pilots and air officers which this country had created. They were filled with enthusiasm, with the full knowledge that air power was the dominating factor in the world's development, and with a perfect willingness on their part to give up their lives to demonstrating its usefulness and to bringing this great, new development to the point that would make America the world's leader in aviation. In 1919 we laid out a plan of development, which, although delayed at times, has been and is being followed at the present time.

The theory is to show that aeronautics can establish airways anywhere in the world and be able to operate from them; that wherever air power can operate, it can dominate sea areas against navies, and land areas against armies; that aircraft can establish the most rapid communication ever known between all of the great centers of population of the world and to the most remote and inaccessible points. This would give to all people, no matter how far removed from civilization, the benefits of rapid communication and the services that go with it. It was and still is a hard problem but the strides towards its solution by the American Air Service since 1919 have been tremendous.

Our first practical trial of these ideas came in the summer of 1919. We organized a definite airway across the continent of the United States, from New York to San Francisco. This had airdromes every two hundred miles, with intermediate landing fields every fifty miles, all of them connected by telegraph, telephone, and some by radio. We had a weather service installed and had gasoline, oil, mechanics, and spare parts distributed to all airdromes. After we actually began work, it took us only two weeks to put in this airway. The arrangements were as complete as was necessary for any operation, even with what we know now, with the exception of night lighting for the airway, which had not been developed at that time. Our airplanes, also, were not suitable for night work.

About thirty airplanes started from San Francisco and the same number from New York, to go across the continent and back as rapidly as possible. The interest was so great in this contest that prizes were offered by many people to the pilots making the fastest time. It was a test of whether airways could be installed and whether pilots could find their way across the country with the maps we had, and whether the engines could stand the continuous running required. The pilots were started on their way and success immediately attended the trial. The airplanes surmounted all difficulties; they traversed the mountains, the forests, the river valleys, the deserts, and all the obstacles intervening. They landed at sea level; they landed on the high airdromes in Wyoming and Utah, several thousand feet above ocean's floor. The leading planes all made about the same speed, that is, about twenty-four hours flying time each way, between New York and San Francisco. Lieutenant Maynard, an ex-Methodist minister, won the contest, with the others close behind him. He was a picturesque character who did a great deal to develop long distance flying. This transcontinental test was an absolute and complete demonstration that air power could be handled over great distances with comparatively little effort; that the airplanes and engines that we had, even then, were capable of long continuous service; and that our flying personnel was perfectly capable of finding their way, anywhere, in any country, under practically any conditions.

The control of the whole undertaking was perfect—the starting of the airplanes, the reports when they passed over localities, and their refueling and care at the airdromes. The test led directly to the establishment of the air mail service between New York and San Francisco, which is the best air service carried on anywhere in the world.

There were several accidents in this contest which

were directly traceable to lack of experience on the part of the pilots. Almost all of the accidents were among officers that had remained in the United States and had not had war training, and were, therefore, not so accustomed to flying across country in unusual and unfamiliar places as were the war pilots. This was immediately corrected so that thereafter all officers were required to fly across country and into all of the states of the Union.

The continental speed tests of 1919 really marked the beginning of the leadership of America in the application of air power, that is, in demonstrating how air power could actually be used, which has continued up to the present time. Our accomplishments have borne a great deal of fruit. They have been watched by the foreign nations even more carefully than they have been by our own Government, and the lessons from them have been carefully digested and methods adopted to use the advantageous features. In this way we have paid, to a very great extent, for the pioneering and experimentation, while others and the whole world have derived benefit.

During all of this time we were working on our bombs and equipment for sinking battleships. Try as we might, we could not get battleship targets from either the War or Navy Department in either the year 1919 or 1920. We gave intensive training, however, to our pilots in bombing and worked hard on our bomb sights, methods of navigation and tactics for this purpose.

Having demonstrated that we could go across the

United States, we wanted to demonstrate that we could establish an airway to Alaska and Asia. The chance came in 1920, when, with the assistance of the Canadian authorities, we established an airway from New York through Canada to Alaska and Nome. Captain Streett took four airplanes from New York to Nome in the flying time of fifty-four hours, and returned in approximately the same time, without the loss of an airplane or a man. The airway again worked perfectly. At Nome, Captain Streett's men stood on the threshold of Asia and could have crossed to Siberia in an hour and a half's time. The reason that they did not was that we had no diplomatic relations with Russia and our State Department did not wish us to land in Russian territory. At that time, even, we could have made the circuit of the globe had we been allowed to prepare for it.

There was now no doubt that airways could be established anywhere desired in the world, and that not only military but also civil and commercial aviation could be used along them. The only question that remained was, to find out what it would be economical to use civil aviation for; in other words, what the relative cost between transportation by air, transportation on the ground and on the water would be and what articles could be transported with profit. In a military way, as to what actually could be accomplished from these airways in the attack of land and water establishments.

In the meantime, we had tried hard to get one of the largest Zeppelins in Europe. We had sent Colonel

Hensley to Germany to obtain one. The contract had been concluded, the money sent over, and the work started by the Zeppelin people on a ship for this country, when the work was stopped and the whole proposition halted for several years. This was the beginning of the acquisition of the Zeppelin ZR-3 which has just been delivered to the United States. We wished to use the great airship for reconnaissance problems, for transporting goods, supplies, and men, for fighting other airships; and as an airplane carrier, that is, equip it with airplanes which could fly away from it and return to it, so that it could go any place over the ocean, launch the airplanes, let them do what was desired, and then have them return. The landing of airplanes on airships is not a difficult problem, as they can fly up under the airship while it is going at full speed and hook on to it. Over land it could do the same thing. The airship, or lighter-than-air dirigible, has the greatest cruising radius of any known means of transportation. This, of course, includes seacraft on water as well as any means of communication on the land or in the air.

We saw that our armament was rapidly becoming sufficiently perfected to destroy the greatest battleships. The sinking of ordinary merchant vessels, torpedo boats, or destroyers and cruisers, is comparatively easy, but to destroy the great battleships is a different matter. These war vessels have been developed consistently from the days of the old galleys: First, they were heavily armored against attack from missile-throwing weapons and cannon against their sides; next, the decks were armored for protection from plunging fire falling from great heights against them; then, after the advent of the torpedo, their sides were honey-combed so that any wound against them under the water line would be localized and would not affect the whole ship and cause her to sink.

It had always been held as a principle that the vessel had to be hit directly by a thrown missile in order to affect it. In studying the battleship, we found that its bottom was its most vulnerable place. It contained no armor and had sticking out of it the open water pipe connecting with the condenser system. The use of the water hammer, or water impelled with great force by an explosion under the bottom of the vessel would certainly cave in the bottom, spring the seams, and cause the vessel to sink. Most of us remember how, when boys, we knocked two stones together when diving under water and what an effect this had on our ears. It is this force that we utilize in attacking ships in the water. Explosions deep in the water at a distance from the ship would break the condenser system, which would stop the vessel from steaming. In fact, one of the most vulnerable points in a battleship is the condenser system. The propellers and their shafts could be bent, the rudder damaged, and the whole under water integrity of the vessel would be deranged. So we determined the best depths in the water at which bombs should explode to get these effects and made our fuses accordingly. These tremendous missiles, containing upwards of one thousand pounds of TNT, could not be tested in any ordinary way. We had to try them in water of various depths, taking note where they exploded merely by looking at them, because any instruments, nets, or wires designed to tell at what depth detonations took place, would be blown entirely to pieces and show nothing. We carried out the experiments in the water in the upper part of Chesapeake Bay near Aberdeen, Maryland. These terrible detonations killed thousands of fish, tore up the bottom, and stopped traffic in the vicinity while the experiments went on. The fuses of these great bombs were also arranged so that if they actually hit a vessel on its deck, they would cause an explosion which would dish or crack the deck, smash up the superstructure, tear down the masts, kill all exposed personnel by the detonation and others by the concussion, put out of commission the telephones, electric light systems, and speaking tubes, and would probably blow up the magazines and the boilers. One very serious difficulty in a modern battleship is its weight, due to its armor, which makes it quite top-heavy and easily unbalanced and sunk by the explosion of a bomb close to it under the water line.

Our experiments had gone forward so far in the fall of 1920 that I was able to announce definitely to Congress that we could destroy, put out of commission, and sink any battleship in existence or any that could be built. This resulted in an interesting controversy that merely showed how any innovation, particularly in methods of warfare, is kept down by the more conservative elements. It is an amusing fact that the Secretary of the Navy at that time announced that these things could not be done and that he was willing to stand on the bridge of the ship while we bombed it. Congress, however, took another view and Mr. Anthony in the House of Representatives, and Senator New in the Senate, introduced resolutions authorizing the President of the United States to designate warships to be used as targets for our experiments. Fortunately, the German prizes of war were about to be turned over to the United States, and, under the conditions, they had to be destroyed within a certain time, the idea being to let the various allied countries learn all the lessons they could from them, then sink them so as not to add them to their naval strength.

Forced to action by the joint resolutions of Congress, the Navy Department began drawing up conditions under which these vessels should be destroyed. There were several classes of ships: Submarines, destroyers, a beautiful cruiser, the Frankfurt, and the dreadnaught Ostfriesland. This splendid ship was designed under the German Admiral Von Tirpitz' orders especially for use in the North Sea against England, where many mines and torpedo attacks could be expected. Her under water construction was the best known and remains a good pattern to this day. Her water-tight compartments were many and each one had solid bulkheads or partitions not even pierced by doors, so that there would be no chances of there being openings in them in case one or more were flooded. She had three skins to her bottom and, of course, was very heavily armored. She was called the "unsinkable ship."

She had participated in the battle of Jutland, had been hit by many projectiles, among them some of large caliber, and in addition two mines had hit her below the water line. In spite of this, she had made harbor under her own steam and had been thoroughly repaired. She was indeed a hard nut for us to crack on our initial attempt at something that had never been tried before. We had meetings between naval officers and air officers to arrange a program of bombing so that the maximum amount of knowledge could be gained from the experiments. The navy insisted on anchoring these target ships on what is known as the one hundred fathom curve which lies about seventy-five miles out to sea from the mouth of Chesapeake Bay.

The airdrome at Langley Field, where the planes were to be concentrated for the bombing maneuvers, was twenty-five miles away from the mouth of Chesapeake Bay, which, added to the distance out from the capes where the ships were to be, made it about one hundred miles over the water and, as we might have to maneuver for an hour in approaching the targets and discharging our projectiles at a speed of one hundred miles an hour, we would probably have to fly three hundred miles over the water on each bombing raid. It is perfectly practicable to go over the water, but any accident to a plane necessitating its landing in the water, might result seriously. In time of war, of course, we would take the chance; in time of peace, it was entirely unnecessary to send airplanes so far out at sea.

There were two other places that could have been

used, one at Cape Hatteras where the hundred fathom curve came within twenty miles of shore; and the other at Cape Cod, where it came within ten miles of land. But the majority of naval officers were so sure that the air attacks would prove ineffectual, that it was desired to show as many Congressmen as possible how little could be done by the air force, and as the sea off the Chesapeake was the best place for this, it was chosen.

The reason that the ships were anchored in one hundred fathoms was because it was required that they be sunk in deep water by the terms of the international agreement, and, next, our bombs would not have so great an effect as they would if the water were shallow. In shallow water, if we can burst the bombs against the bottom, they explode up and against the bottoms of the vessels with very much greater force than where the water is deeper, in which case there is no "tamp" upward against shipping.

It was up to us, however, to show that what we advocated could be done, so we had to accept the conditions as they were offered. These made us operate under conditions that were about as hard as could be drawn up for the accomplishment of the difficult experiment.

Meanwhile, orders had been issued all over the United States for the concentration of our pilots and airplanes at Langley Field, Virginia. This airdrome lies only a short way from Hampton, which is one of the oldest, continuously inhabited, towns in America. It lies not far from where General McClellan's army, equipped with the first military aircraft in the form of Professor Lowe's balloons, invested Yorktown in 1862; where Cornwallis surrendered to General Washington and Lafayette; and only a few miles from Jamestown where John Smith planted his colony; and from where the Monitor and Merrimac had their great struggle. The results of the bombing might hold more far reaching consequences than even those events. Our pilots, observers, and mechanics had their imaginations fired with these events in our history, and were determined that the maximum results should be accomplished by our little air force.

The airplanes started coming in from the West, from the North and from the South. Three heavy bombers —one Handley-Page and two Capronis—made the flight from Texas. It was the first time that a continuous flight had been made by these large ships for such a distance. With them came their veteran crews of the Border Patrol.

What had remained of our air force had been deployed along the Mexican Border in 1919 and 1920, so as to protect our frontier in case of trouble. Our Martin bombers, large two-engined airplanes, had never been tried out in service for they were just being finished in the Martin airplane factory at Cleveland, Ohio. Our experience in flying large bombers from Italy to the western front in Europe had been quite disastrous, as nearly all crashed in transit across the Alps, and great fears were entertained that many accidents would happen to these great ships on their long journey. The planes were so well built, however, the crews so expert, that not one of the thirty transported was destroyed en route.

As each airplane came in it was assigned to its particular organization. The staff was organized along the same lines that we had found to be the best in Europe and regular plans were drawn up for the whole organization and for every operation. The air force was known as the First Provisional Air Brigade and it had every element necessary for the operation of a large air force. A pursuit aviation under Captain Baucom, who had so greatly distinguished himself in Europe, was organized to act as protection to the great bombers in case they were attacked by hostile pursuit aircraft. The light weight bombers, for which we used the old De Haviland airplanes, were designed to attack torpedo craft, transports, and light vessels, and, in case the anti-aircraft artillery from the warships amounted to anything serious, they could attack the big ships with their light bombs and machine guns at close quarters so as to nullify any effect against the heavy bombers from this source.

The heavy Martin bombers with their cruising ability of five hundred and fifty miles and their power to lift from two to three thousand pounds of bombs, were the most powerful airplanes that we had ever had. At first our men were unaccustomed to them and were a little nervous when flying them with their full loads, but they soon found how very airworthy they were and gained the utmost confidence in them.

The crews were first trained carefully over the land and then began working over the water. Both still and moving targets were used and, in some cases, automobiles were run along the roads to indicate the maximum speed at which seacraft might be able to go on water. Corners were turned and all changes of course in dodging on the part of the war vessels were simulated. Each class of seacraft was carefully studied and all of the personnel made familiar not only with their appearance, but with their structure and the sizes of bombs that were necessary to destroy each class of boat.

We were allowed to use only one weapon in the first maneuvers against the ships anchored off the Virginia capes. These were our high explosive bombs. Air forces, of course, can use contact mines, gas and smoke screens of various kinds, phosphorus bombs, thermite to cause great heat, water torpedoes, air torpedoes, and gliding bombs. The use of these weapons was explained and taught to our officers.

All of these airmen had been flying from three to five years. They excelled any personnel ever brought together or probably that ever will be brought together, in their determination, their ability, their confidence in their organization, and their knowledge of air tactics. Fortunately, the concentration was effected at Langley Field without the loss of a single pilot and, in spite of the fact that these great weapons were being used constantly, comparatively few accidents actually happened.

One bad one occurred at Aberdeen Proving Grounds in the upper part of Chesapeake Bay, with quite a small bomb. The bomb had been hung on the wing of a DH airplane preparatory to taking it into the air for a shot at a target. The bombs' primers or fuses are so arranged that they are not set or "armed" as we call it, until they are dropped from the airplane. When the bomb drops a little wire releases a small propeller which is on the nose of the bomb. The pressure of the air makes this turn which actuates a screw that releases a plunger which acts as a firing pin against a detonator or cap. Some bombs have this little fly wheel in both the nose and the tail so that if one fuse fails to work the other one will. In the case of the bomb at Aberdeen, the mechanism for dropping the bomb released it while the airplane's propeller was being turned upon the ground preparatory for flying. As the bomb lay on the ground, the stream of air rushing from the propeller turned the arming device over and made the bomb "live." Not seeing it, the pilot started to taxi the airplane out on the airdrome and, in turning, hit the bomb with the tail skid of his plane. A terrific explosion immediately followed. It killed three men and wounded thirteen or fourteen. As this was only a fifty pound bomb it can be imagined what would have happened if one of the larger sized bombs had detonated. We safeguard all of our weapons in the greatest way possible and, in order to have the bomb explode, four different things have to happen. In this case, all of them did happen: first, it dropped off; second, it had a stream of air against it so as to actuate the little propeller on the bomb; next, the plunger was caused to be pressed against the detonator; and, last, the bomb was hit with a severe jolt by the tail skid of the airplane. It is remarkable that more accidents do not happen when man attempts to harness these great elemental forces.

Within a couple of months before the coming of the warships the organization of the Brigade at Langley Field was complete. Practice was begun over the water. First, we drew out a silhouette of a battleship on a marshy point near the mouth of the Back River, marking the exact points where the bombs should hit. It was practiced on every day. Then we obtained a tug and towed a target representing a battleship out into the Chesapeake. The bombing became so accurate that the crew of the tug had no fear in using a line only about three hundred feet long. Every day direct hits were made on these structures with our one hundred pound, sand loaded bombs, so that it was hard to keep sufficient targets on hand. Next, we bombed the wrecks of the old battleships, Texas and Indiana, close to Tangier Island, about fifty-four miles away from the airdrome. On these we used live bombs of heavy weight.

There, in the haze of the morning and evening, we ran into quite unusual conditions. As we were about eight miles away from the nearest land, it was often invisible and we found ourselves with no horizon or point of reference by which to level our planes, because everything was the same color—air, sky, and water. It was very much as if we were in the inside of a sphere all painted the same hue. Many who had had experience in over-water flying had held that it was impossible to level an airplane so that good bombing could be done for this reason, but we soon found by the aid of a gyroscope brought to us by Lawrence Sperry of the Sperry Aircraft Company, that we had no difficulty in maintaining direction or flying the planes on a level keel. One learns these things only by actual air work and experience. With the comparatively crude bomb sights that we had, this was a most important consideration. By that time our bombing, on all sorts of targets, was becoming so accurate that even the most doubting of our officers knew that whether hostile seacraft were at rest or moving, no matter how fast, there would be no difficulty in hitting them.

We then practiced at night. The airplanes left in formations at night, went through their regular bombing practice in the same way they did in the daytime, searched for their targets out over the water, signalled to each other and made their attacks. Each one of the Martin bombers was provided with radio telephones so that one plane could talk to the other. Flights were now made up and down the coast and every one was made familiar with the distinguishing marks on the various lighthouses and life-saving stations, and the places where suitable landings could be made in case of emergency.

As the time became shorter before the arrival of the target ships, the excitement grew.

Our bombs began to come in—the three hundred pounders, the six hundred pounders, and, last, the eleven hundred pounders. We had had some concrete bombs made the same size, shape and weight as the two thousand pounds. With these we had tested the bomb carriers on the airplanes and the mechanism for dropping them, as to insure their proper functioning. We took up a few of the bombs and dropped them to see if everything worked satisfactorily. Not one of these bombs failed us in any way. Their shapes made them fall straighter than any bombs we had ever seen used in Europe or in this country, and their fuses were the best we had ever seen. Still, our two thousand pound bombs did not arrive and as the months shortened into weeks and the weeks into days before the tests were to take place, we grew a little nervous at their non-delivery. At last we found that these monster weapons were on their way and learned one cause why they had been so late. It seems that the steel bomb cases had barely been finished two weeks before the test. It took several days to install the fuses and then the half ton of TNT had to be poured into each one of them in a molten condition. No one had known before that it took such great masses of TNT a very long time to cool. In the ordinary summer temperature it would have taken ten days or two weeks before they would have been ready for shipment, in which case they would have been too late, so, to overcome this, our resourceful Ordnance Department packed each one in ice so that within a couple of days they were on their way. They reached us just in the nick of time and were stored in our ammunition dumps with sand bags piled around them to localize any explosion in case one went off. This would have done little good, however, because, if a single one had detonated, probably the whole airdrome with all of its equipment would have been destroyed together with most of the personnel at that place.

The First Provisional Air Brigade was now ready to attack any warships or fleet of warships which could have been sent against that part of the coast.

The personnel was experienced with all of the difficulties incident to using the great bombs and knew how to obtain the best results from them. We also had the best planes that we could get hold of, but they were not at all suitable for work at sea, as they had been designed for work in Europe. Some of the little pursuit planes had only sufficient gas to get out to the target battleships and back again, but their pilots were determined to go even though they might have to land in the sea and be rescued by one of the patrol vessels. The First Provisional Air Brigade was also supplied with a squadron of seaplanes, equipped for rescue purposes, with doctors, first aid equipment, and means for carrying patients. These, however, did not prove very successful because seaplanes, when required to land in heavy waves, usually smash in their bottoms and often sink more quickly than the ordinary airplane buoyed up by the air in its wings. The seaplane squadron was carefully trained in the work for which it was intended, and whenever it had smooth water to land in, did good work.

In our large lighter-than-air hangar we had four small dirigible airships. These had the ability of staying in the air from twenty-four to thirty hours and of making a speed of about sixty miles per hour. They could operate by night or by day. They, also, were trained in rescue work so that if forced landings in water occurred, they were able to go down near the water, drop rope ladders down to the swimmers, and haul them up to safety. This was practiced at different times in the middle of Chesapeake Bay. In addition to their rescue work, they were trained for reconnaissance service, and, because they had very powerful radio telegraph installations on board, they could be heard for long distances, much further than the airplanes. Their use was particularly efficient at night when they themselves could not be seen but when they could easily distinguish shipping on the surface of the water, either by seeing the ship directly or by dropping the large calcium flares which produce lights of several hundred thousand candle power, that illuminate great stretches of the ocean. Ordinarily, they had little trouble at night in picking up seacraft on account of the phosphorescent wake and the white waves thrown up by the bows and sterns of the vessels.

We had photographic planes equipped with both still and motion picture cameras so as to make a complete record of every shot. We had a completely organized meteorological service and the means of transmitting weather information to all of our airplane squadrons and of predicting weather from twenty-four to thirty hours ahead of time. Cloud flying, rain flying, and flying in storms, were practiced and well understood by all.

The Chesapeake Bay region is subject to terrific thunder storms during the summer time. These are caused by the heat of the sun making ascending currents of moistened air which gradually form the cirro-cumulus clouds. These work inland and toward the afternoon they meet the cold air currents that come down the sides of the Alleghany Mountains. This condenses the moisture, forming heavy clouds which, in turn, produce "thunder heads" which gradually sweep down over the bay regions in the afternoon with ever increasing speed accompanied by discharges of lightning and heavy rains. Our instruments have registered velocities of over one hundred and twenty-five miles an hour on the ground in these storms. We had to be particularly careful in handling our air organizations so as to get them through and away from these storms while operating out at sea. The average storm in this region is not over twenty-five miles in diameter nor does it move across country at more than twenty to thirty miles an hour, but, inside it, the wind velocities are very high. Near the ground the wind gets as if it were a cylinder of air that rolls along the earth. This is caused by the air hitting the ground and being retarded in very much the same way that the water at the bottom of a river is retarded when it touches the earth so that the water at the bottom of a river is always flowing more slowly than that at the top. With the air, however, the retardation at the bottom allows the air above to pour or slip completely over until it touches the ground. The result is that we may get up and down currents that are so severe, during heavy storms, as to dash airplanes against the earth before they can pull out with their engines.

Another interesting thing that we found out was

that if we flew within eight or ten feet of the water we obtained more life and, consequently, more speed. This is due to the fact that the wings of an airplane push the air downward and pile it up. When this air hits a flat surface underneath, it acts very much like a cushion under the plane and gives it greater sustentation. We use this greatly to our advantage when going against heavy winds, because by getting down close to the ground the winds lose a great deal of their velocity, whereas we in the airplanes gain life or sustentation and speed.

At last the day arrived when the target warships were put into position and we were given orders to stand by for the first attack. We were now to be given a chance to prove whether aircraft could actually sink the latest types of battleships. If this were successful, it would mean eventually that aircraft would control all traffic on the seven seas and that, as they would eventually be able to destroy and attack all communications on land, the outcome of these maneuvers would cause an entire rearrangement of the elements of national defense which each country possessed at that time.

## III

## THE UNITED STATES AIR FORCE PROVES THAT AIRCRAFT DOMINATE SEACRAFT

THE stately Atlantic fleet, consisting of eight battleships, several cruisers, many destroyers and auxiliary vessels, hospital ships and tenders, moved into the Chesapeake Bay and anchored in the Lynnhaven Roads. The appearance of these great vessels was majestic. The fleet had been assembled to watch and observe the bombing tests, so that all could see what happened. Many considered this trial to be utterly useless, because they reasoned that it was entirely impossible to sink, or even injure, a battleship. That, neither could a battleship be hit by an aerial bomb, and, if it were hit, could it be damaged to any great extent. These people little knew the great accuracy of aerial bombing, which at the present time, at battle ranges, is perhaps the most accurate method of hurling missiles. Also on account of the tremendous proportional amount of explosives carried in the bombs they constitute the most powerful weapon of their kind ever devised by man.

On the other hand, the officers and men of the First Provisional Air Brigade were certain that their efforts would be crowned with success. They were eager to have their theories put to the test.

The first trial took place on June 2, 1921, against the ex-German submarine U-117. It was anchored on the target grounds, about seventy-five miles off the Capes, in an awash condition. The day was perfect and the line of the destroyers placed at ten mile intervals across the Chesapeake Bay and from the Capes out to the target ships, looked like little beetles in the molten sea as we sped over them.

The first air units to attack this submarine were three flying boats of the Naval Air Service, commanded by Lieutenant Thomas. They flew close together in a "V" shaped formation, and each dropped one bomb for sighting purposes, as they passed over the target. The practice was perfect and each missile either hit the submarine or struck within a few feet of it. Making a turn, they came back dropping three bombs each, or a total of nine bombs. These bombs weighed one hundred and eighty pounds each. The center of impact of this volley struck the submarine squarely, split her in two and down she went.

None, except the air people, had expected such a rapid termination of the first experience. Had she been fired at with cannon no such result could have been obtained—she would have been hit above the water line and would have had to fill up with water gradually before she would have gone down. These bombs tore her all to pieces above the water, below the water, and along the water line. Some of the skeptics began to be convinced that there was something to air bombing.

It was a very severe jolt for those who still adhered to the old theories.

This was not all: The vessel that was directing the target practice, which remained about a mile and a half away from the submarine, had her condenser system so badly damaged that she steamed into the Chesapeake Bay that night at only three knots an hour. If these small bombs could produce such an under-water shock as to affect a warship's condenser system a mile and a half off, what could the big bombs do?

The discussions waxed stronger than ever, and the Congressmen, naval officers, press representatives, and guests on the Naval Transport Henderson spent the night in terrific arguments. These arguments, however, were soon to have their answer.

A couple of days later, the Army Air Service was given as a target the "ex-German Torpedo Destroyer G-102." She was one of the large type boats of this class, employed so successfully by the Germans. Our Air Brigade moved out with all its different parts in exactly the same way that it would have attacked any sea force, equipped with airplane carriers and airplanes. First came the pursuit ships, eighteen of them, flying in three flights, so that they could attack, underneath, up above, or on the same level, any enemy pursuit ships which might contest their progress. They not only had their machine guns, but each plane carried four twenty-five pound bombs to be used against the ship's superstructure, anti-aircraft guns, searchlights, and personnel, so as to sweep the deck clear and interfere with her navigation. Should one of these small bombs be dropped down the funnel into the boiler rooms and detonate, it would cause the boiler to explode, and, of course, blow up the vessel.

The pursuit pilots thought they would be able to sink this vessel alone with their twenty-five pound bombs, but I hesitated a long time before allowing them to go so far to sea, as their old airplanes had only a two hours supply of gas. They insisted on being allowed to go, however, so that they could carry out their allotted part in the maneuvers.

Following the pursuit aviation, went the light weight bombardment squadrons of DH two seaters, each plane equipped with four one hundred pound bombs. These bombs alone were sufficient to sink a destroyer or any unarmored ship, and in case a battleship was attacked, they could at least clear the decks of any exposed personnel, smash up the communicating and radio systems, and probably bend the battleship's propeller shafts and rudders, so that even a heavily armored vessel could not be worked after an attack with these light bombs.

Following the light weight bombers, came the heavy Martins, a dozen of them, each equipped with six three hundred pound bombs. They sped along in their "V" formation about two miles behind the DH squadrons. It was the first time in aeronautical history that an attack had been made in this way. Every element of a large air force was there. First the pursuit aviation to take care of the opposing aviation, and, after having conquered it, to attack with their machine guns and bombs the decks of the war vessels. Next the light weight bombardment to scatter and destroy the auxiliary vessels of the battleships, such as cruisers, destroyers, and submarines, and, last, the heavy weight bombers to sink and destroy the battleships themselves.

We had been allotted this destroyer to attack in any way we saw fit, not with restricted conditions as was the case with the other vessels, so I took this occasion to have the whole air brigade take part so that all could see that the plans that we had worked out were correct, which they proved to be.

I accompanied every bombing mission of the airplanes over the sea in my control ship the "Osprey," a two seater DH plane, with sufficient gas in it to fly for five hundred miles. On that day Lieutenant Johnson accompanied me in a brand new Thomas Morse single seater pursuit airplane. It was able to go about one hundred and seventy miles an hour. I used him as a courier, signalling to him what I wished to be done. He would go and return with the information in a very short time in this very fast plane.

As we approached the target vessel we could see the whole Atlantic Fleet formed in a circle around it. We wound our way in and out of great banks of cirro-cumulus clouds behind which we could have concealed our approach very easily had it been actual war. As the pursuit squadron arrived over the target, they formed into a large "V," preparatory to attacking with bombs, as it was assumed that the hostile pursuit aviation had been defeated in the air battle. Captain Baucom, leading them, gave the signal for the attack and one after another in regular cadence dove for the destroyer, coming straight out of the heavens for three thousand feet. Straight toward the destroyer they went until within two hundred feet of it when they let go one of their bombs. The airplanes followed each other at about thirty second intervals, so that there was a continuous stream of bombs against the target. It had been decided beforehand that a certain number of bombs should be fired at the decks, a few exploded in the water alongside, and some hits made down the funnels in the expectation that these small bombs alone might sink the vessel. The attack was beautiful to watch; the accuracy of the bombing was remarkable, practically every bomb went where it was directed. The decks of the destroyer were punctured and swept from end to end. Lieutenant Alsworth put one bomb straight down one of the funnels, which undoubtedly would have blown up the boilers. Every one was surprised at the great accuracy of this bombing.

Under cover of the pursuit attack the light weight bombardment airplanes with their hundred pound bombs moved up. The method for attack found to be the best was to fly one airplane behind the other at a distance of about two hundred yards, so as to bring a succession of bombs falling at the target with only a few seconds interval between them. In this way, one airplane could correct from the other's fire in front. Also, it would be impossible for a vessel to escape by changing its course or dodging, because it would be seen from the airplane succeeding in the column and its fire would be corrected accordingly. We had taken up so much time in the pursuit attack that I feared the fuel would run short in the two seaters, so I called off this attack and ordered the Martin bombers to finish the destroyer. They moved up at once, directly into the wind, led by Captain Lawson, who so greatly distinguished himself in all of the bombing. Behind him stretched the squadron of twelve great airplanes. In less time than it takes to tell it their bombs began churning the water around the destroyer. They hit close in front of it, behind it, opposite its side and directly in its center. Columns of water rose for hundreds of feet into the air. For a few moments the vessel looked as if it was on fire, smoke came out of its funnels and vapors along its decks. Then it broke completely in two in the middle and sank down out of sight.

The demonstration was absolutely conclusive. While it was not particularly difficult to sink the vessel itself, those who thought any anti-aircraft guns could keep off an air attack, saw that it was now impossible because under cover of pursuit and light weight bombardment aviation, the larger bombers could move in with little danger. All our methods and systems of bombing had proven to be correct. The bombs themselves never failed to explode. The accuracy was remarkable and the spirits of every man ran high. We had, however, stretched our cruising ability out at sea to the limit with some of the old airplanes that we possessed. Many of the planes of the pursuit squadron barely got back to the shores of the coast where they landed along the beach. All the airplanes we had except the Martin bombers were obsolescent war machines entirely unsuited for this work. There were no injuries and no forced landing in the water, and due to the excellence of our mechanics very little trouble was experienced with the engines. This again surprised the onlookers, who had expected the airplanes to fall into the water and that disaster might attend our efforts.

One Martin bomber barely made the shore. Lieutenant Dunlap was the pilot. One engine failed about fifty miles at sea, so he made for the nearest point on the shore. The bombers, it will be remembered, have two engines, and with the other engine he held up the ship as best he could, constantly losing altitude as one engine alone could not keep the bomber in the air. He just made the beach and landed safely. His one engine, however, had become so hot that when he cut the spark it would not stop and, as the engine in the bombers is set off on one wing, it turned the airplane around and dove it right into the breakers along the beach, where the large waves broke it up. Lieutenant Johnson, who had acted as my courier in the small plane, had also run out of fuel and he had made a landing near the spot where the bomber had met with misfortune. It would have been very difficult to send more gas from Langley Field for Johnson, a distance of thirty-five miles, so he swam out to the Martin bomber, now being pounded to pieces in the surf, cut the emergency tank of gasoline out of the upper wing, poured the gasoline into his own airplane and flew to Langley Field.

That night all of our men had returned safely to Langley Field after their first great experience in bombing. The rejoicing was tremendous. They knew now that unless something most unusual happened, it would be proven for all time that aircraft dominated seacraft.

The next operation scheduled was to search for and find a battleship that was supposed to be located anywhere between the mouth of the Delaware River and the mouth of the Chesapeake Bay, and then to bomb it with sand loaded bombs. It was to be under steam and controlled by radio telegraphy. Its speed was only six knots an hour under these circumstances. This was such a simple problem that there was no use of sending any airplanes out over the water for that purpose, particularly as we had so few of them and as this test would give us very little practical benefit. So some of the dirigible airships were sent out and promptly found the ship, which was the Iowa, and sent back news of its whereabouts. Seacraft are not only very easy to find, but their type and character are also as easy to determine from the air.

The air bombs had now sunk the unarmored ships, that is, the submarine and destroyer. These tests did show conclusively that planes could sink merchantmen, transports, or any kind of a vessel not protected by armor.

Our next target was to be the cruiser Frankfurt, a beautiful vessel. It had considerable side-armor, deck-armor, excellent watertight compartments and bulkheads, and every perfection of a modern vessel of that class. The tests were to be conducted with varying sizes of bombs and after each attack with a specified number of bombs which was not intended to sink the ship, an inspection would be made by officers of the navy to see what damage had been inflicted. These tests started on July 18th, and, of course, the interest grew as they went on. First one hundred pound bombs were tried, and then three hundred pound bombs. The three hundred pound bombs, undoubtedly, would have sunk the ship had not this part of the test been called off by the navy. The vessel was then thoroughly inspected and the damage resulting noted. At last came our chance to attack the cruiser with the six hundred pound bombs and again Captain Lawson led his squadron to the ship. At that time the Board of Inspection was so slow that they kept us flying around way out at sea about an hour before Captain Lawson had to signal that unless he was allowed to attack within fifteen minutes, he would have to return to shore on account of lack of gas. At last came the order to go ahead. Captain Lawson deployed his bombers into single column and immediately went for the target. The bombs fell so fast that the attack could not be stopped before mortal damage had been done to the ship. The control vessel made the signal to cease bombing as the good ship was toppling over, so quick was the effect of the bombs.

Many amusing things occurred. At the first direct hit of a bomb on the Frankfurt's deck, fragments of steel were thrown over the water for over a mile. The crews of the observing battleships had crowded to the rails to watch but as these pieces of steel came nearer and nearer to them, they rushed to the other side of the vessel for protection. It made one think what might happen in case a real attack was made against naval vessels in war, whether the crews could be held to their posts in view of almost certain destruction. As Captain Lawson's bombs fell tremendous columns of water shot up. Some fell in tons on the deck of the ship, sweeping it clear. It was the first time we had used our six hundred pound bombs at sea, and they worked splendidly. From the time the cruiser Frankfurt received her mortal blow, she sunk rapidly toward the port side, then slid down bow on. She was soon out of sight and again it was proven that our air bombs could destroy a cruiser as no other weapons could.

While the bombing was proceeding during the day, before the final attack, when only a few bombs of small size were allowed to be fired and the bombing stopped instantly when it looked as if any damage had resulted, the naval contingents began to feel that this vessel might resist the air attack and arrangements were all made to sink it by cannon fire from one of the battleships so as to demonstrate how quickly cannon would be able to put a vessel of this kind under water. As was to be shown later, cannon fire had very little effect compared to aerial bombs against these floating hulks lying in the water without steam in their boilers and without ammunition in their magazines. The cannon were able to make holes above the water lines mostly, whereas the bombs blew in their bottoms.

At last the time came for the bombing of the Ostfriesland. This was our real test. If we could not sink this great ship the efforts against the other smaller vessels would be minimized and the development of air power against shipping might be arrested, at least for the time being. No foreign air service had been able to obtain battleships as targets, as such action had always been strenuously opposed by their navies. Ours was the first to get them through an Act of our Congress. About the fifteenth of July we began the test, firing bombs of small calibre against the Ostfriesland. This bent up the equipment on her decks and caused some other damage, enough to put her out of business but not to sink her. We knew full well that the very large bombs, eleven hundred pounders and two thousand pounders, would be necessary to sink her and that the little bombs, from our standpoint, were very largely a waste of time. We had to kill, lay out and bury this great ship in order that our people could appreciate what tremendous power the air held over battleships.

At last we were allowed to take out our eleven hundred pound bombs on the 20th of July. We, however, were ordered to drop only one of these at a time, instead of two at a time. An impact of two of these in any place near the ship would probably have sunk her. It was desired, however, to observe the effect without sinking the ship. This attack was made by a flight commanded by Lieutenant Bissell, and was a perfect exhibition of airmanship. Earlier in the day we at Langley Field, a hundred miles away, had heard nothing from the fleet and as the hours went on and it became time for our attack, and, as our crew were waiting beside the loaded airplanes for the order to go, still no word came. Captain Streett, my observer, and I proceeded out over the ocean to see what the trouble was, and to our astonishment we saw the whole Atlantic Fleet making for the Chesapeake Bay. At first we could not understand why this was but later we found out that since there had been about a twenty knot wind blowing, they determined an airplane could not act. Some said that many on the warships were becoming seasick. We then signalled that we wished to begin at once, whereupon the fleet returned to the targets. Aircraft can act in high winds and under weather conditions where seacraft have great difficulty in doing anything. A twenty or thirty mile wind is a great difficulty on water, whereas it amounts to very little in the air. A wind double that strength can be flown in whereas with shipping, strenuous measures have to be taken for safety.

Arrived at the target Lieutenant Bissell's flight of five planes deployed into column and fired five bombs in extremely rapid succession, in fact, it looked as if two or three bombs were in the air at the same time. Two of these bombs hit alongside and three hit on the deck or on the sides, causing terrific detonations and serious damage. Fragments of the battleship were blown out to great distances. Spouts of water hundreds of feet high shot into the air. We felt the jolts and noise of the explosives in the air in our planes three thousand feet above where the bombs hit.

Immediately the Navy Control Vessel made frantic signals for the attack to stop. Lieutenant Bissell had turned his flight and was ready to finish her, as he had five additional bombs left. He had injured the ship so severely that if she had been equipped with her crew, her ammunition, and had had steam in her boilers, she probably would have been destroyed. That night she listed so badly that two thousand tons of water were let in on the other side to keep her straight up so that she would not roll over. Just as Bissell's attack was ceasing, we saw a storm driving in from the north. We had received no intimation of this from our weather service. It was a typical severe thunder squall, prevalent at this time of the year in that locality. The dirigible airships that had been watching the maneuvers taking photographs, and ready to rescue any planes which might have to make a forced landing, proceeded to the north at once and escaped. Lieutenant Bissell's flight broke through it toward Langley Field.

In my own plane, Captain Streett and I had waited until all had started back, then we made for the shore. Although we could have gone through the storm, we decided it was better to go around, particularly as I had gathered up some other planes and ordered them to follow. Two of these were large flying boats, and with these it would have been dangerous to go into the storm because they were quite underpowered and unwieldy compared to the landplanes. As the storm was moving to the southeast, and as it was about twenty to thirty miles in diameter, we had a long way to go to get around it and a difficult problem in navigation presented itself. We had to estimate the velocity of the wind, take our direction by compass and our speed from our air speed indicator. Captain Streett, my observer, plotted our course every two minutes, and after we had flown for an hour, I asked him for our location. We were in the edge of the storm, the light was a sort of a dark yellow and there were flashes of lightning all around us. Streett said we were about twenty miles off shore and about ten miles north of the Currituck Lighthouse in North Carolina. Watching my clock I counted twelve minutes, which should have carried us twenty miles, and looking down I saw the beach under us. Instead of being ten miles from the Currituck Light, we were five miles from it. It was a remarkable example of navigation with the instruments we had at that time. Since then we have devised much better ones. Shortly afterwards we crossed Currituck Sound and landed in a cotton field to check up our bearings and planes. We arrived at Langley Field long after dark. The storm was still driving. Our heavy bombardment squadron was lined up on one side of the airdrome and loaded with two thousand pound bombs, ready for the attack the next day. The landing lights on the airdrome did not work very well, and it was a difficult thing to keep our own airplane away from the big bombers loaded with the huge missiles, as we had to land right over them. Had we struck one of these and detonated one bomb, probably the whole of Langley Field would have been destroyed.

On checking up about ten o'clock, I found all machines present or accounted for. In the long detour that we had made no one had been forced to land at sea. Had one done so, he would have been lost as no patrol or rescue vessels were near and the sea was very rough from the storm in addition.

Captain Lawson had had to land some distance south of Norfolk and had broken a wheel. With his usual resourcefulness, he got an automobile, drove to the nearest railroad station and arrived at Norfolk. He crossed the James River and reported to me about one o'clock in the morning.

Bright and early next morning Lawson was ready to take out his squadron loaded with the two thousand pound bombs for the final chapter in our tests. It was now felt we could destroy the Ostfriesland; some thought we should be restrained from doing it because it would lead people to believe that the navy should be entirely scrapped, as a thousand airplanes could be built for the price of one battleship. Others thought it should be done because air power had brought an entirely new element into warfare on the water, and if the United States did not draw the proper lessons from it, other nations would and we would be at a great disadvantage. Those of us in the air knew that we had changed the methods of war and wanted to prove it to the satisfaction of everybody.

Finally the time came for us to attack the Ostfriesland with the two thousand pound bombs, and Captain Lawson's flight went to sea. The great ship was down a little by the stern, drawing about forty feet of water; she had sunk considerably after Bissell's attack on the preceding day. Lawson circled his target once to take a look at her and make sure of his wind and his altitude. He then broke his airplanes from their "V" formation into single column and attacked it. Seven airplanes followed one another. Four bombs hit in rapid succession, close alongside the Ostfriesland. We could see her rise eight or ten feet between the terrific blows from under water. On the fourth shot Captain Streett, sitting in the back seat of my plane, stood up and waving both arms shouted: "She is gone!"

When a death blow has been dealt by a bomb to a vessel, there is no mistaking it. Water can be seen to come up under both sides of the ship, she trembles all over, as if her nerve center had been shattered, and she usually rises in the water, sometimes clear, with her bow or stern. In a minute the Ostfriesland was on her side; in two minutes she was sliding down by the stern and turning over at the same time; in three minutes she was bottom-side up, looking like a gigantic whale, the water oozing out of her seams as she prepared to go down to the bottom, then gradually she went down stern first. In a minute more only the tip of her beak showed above the water. It looked as if her stern had touched the bottom of the sea as she stood there straight up in a hundred fathoms of water to bid a last farewell to all her sister battleships around her.

We had been anxious to sink the submarine and destroyer, but I had felt badly to see as beautiful a ship as the Frankfurt go down. She rode the water like a swan. The Ostfriesland, however, impressed me like a grim old bulldog, with the vicious scars of the battle of Jutland still on her. We wanted to destroy her from the air but when it was actually accomplished, it was a very serious and awesome sight. Some of the spectators on the observing vessels wept, so overwrought were their feelings. I watched her sink from a few feet above her, then I flew my plane above the transport Henderson, where the people who had observed the tests were waving and cheering on the decks and in the rigging.

Contrary to the popular opinion, that great vortices in the water are formed as a ship sinks, there were none in this case. She slid to her last resting place with very little commotion. Thus ended the first great air and battleship test that the world has ever seen. It conclusively proved the ability of aircraft to destroy ships of all classes on the surface of the water.

Later that same summer we were given another battleship to practice on. It was the Alabama. She was towed into about thirty-five feet of water near Tangier Sound in the Chesapeake Bay. Again Captain Lawson's invincible squadron attacked the battleship. This time we went in to sink her as quickly as possible and being in shallow water, the effect of our bombs was greater. The first two thousand pound bomb did its work, she sank to the bottom in thirty seconds. Six other bombers coming behind Lawson struck her with their projectiles and within four minutes she was a tangled mass of wreckage unrecognizable as the fine ship which had been there before. We tried out various weapons against her before she was sunk. Phosphorus bombs gave a magnificent spectacular display, the lapping flames completely enveloping the ship. We put thermite, the greatest producer of heat known, on her decks and covered her with smoke clouds dropped from the airplanes. We attacked her at night and made direct hits with our bombs in the darkness.

Again amusing things happened. One was in connection with our gas attack. Small bombs of tear gas of only one-half strength and weighing twenty-five pounds were dropped on her. They immediately pervaded the whole ship and the officers that went on board to note the effect immediately had to put on their gas masks. Some of these officers that went below became lost as they had neglected to take flash lanterns with them and in the obscurity of the hold, as it was not lighted by electricity, and encumbered by their gas masks, they were unable to find their way around. Eventually they were gotten out after a good scare, as they were afraid that the bombing would begin again while they were down there. The tear gas worked into the clothing of many of the officers so that after they went ashore, hours later, where they were staying in farm houses, the gas exuding from their clothing caused the girls waiting on the tables at dinner to begin to cry.

The vessels observing the tests had to change their anchorage when the wind changed so as to blow the gas away from the hulk of the Alabama to them, although they were nearly a mile away. The tear gas also appeared to be carried up and down by the tide, as it is heavier than air and rested on the water.

The tests against the Alabama in many ways were more interesting than those against the other vessels away out at sea, because so many different weapons were used against her. Any number of weapons may be used against seacraft and these, of course, will be constantly developed from year to year. Chemical weapons, such as gases, phosphorus compounds and acids are enough to put any surface ship out of action.

After looking over the effect of all the different weapons used against the Alabama, one of the officers facetiously remarked, that the future individual equipment of a sailor on a battleship would have to consist of a parachute to come down in when blown up into the air by our bombs, a life preserver to float on the water when he came down. He would have to wear asbestos soled shoes, as the decks would all be hot, a gas mask to protect him from the noxious gases, and a pocket flash lamp to find his way around the deck and interior of the ship when the electric lights went out.

Just as the tests were finished against the Alabama, serious domestic disturbances occurred in the mining regions of West Virgina. This state is extremely mountainous; it is so up and down hill that our pilots reported that the birds and chickens had to have one leg shorter than the other to walk around on the side hills, and there are practically no landing fields in all this area. A reconnaissance of the locality, however, disclosed the fact that we could land almost anywhere in the State with very little preparation.

Immediately after receiving the order, a two-seater squadron under Major Johnson flew over the mountains and landed at Charleston, the capital of West Virginia. The large Martin bombers were used as transport airplanes for this squadron, and carried the medical staff under Major Strong, with their medicines, ammunition for the machine guns, tear gas bombs and explosive bombs. The whole movement was completed in an incredibly short space of time, and neither railroad nor automobile transportation was used in the initial movement. These were the same organizations and same airplanes that had sunk the battleships far out at sea; now they had crossed the Alleghanies and landed in the midst of the mountains. It was an excellent example of the potentialities of air power, that can go wherever there is air, no matter whether they may be over the water or over the land.

At the end of 1921 the American Air Service had conclusively demonstrated what could be done by air forces against seacraft. During the following year, 1922, all the World's flying records were captured by the United States—speed, altitude, long distance, and the hours of time that an airplane could stay aloft. In 1923 and '24 we sank more battleships, flew across the American Continent from daylight to dark, and then our airplanes circled the earth, having established an airway clear around it. Whenever an airway can be established, there aircraft can go. The means of bringing their fuel to them, such as railroads, automobiles, boats or other aircraft are merely their auxiliary means of transportation.

In spite of these splendid performances of individuals, which have led the way for the world in the development of this most important art and science, and benefit to commerce and civilization, we, today, compared to our resources and ability are falling back constantly.

## IV

## CIVIL AND COMMERCIAL AVIATION

TRANSPORTATION is the essence of civilization. The more rapid the intercourse between people, the more highly what we call "civilization" will be developed. Commercial nations have always made it a point to establish and control transportation systems so that their means of distributing their goods might be controlled by themselves and not be dependent on others. No matter how great a producer or manufacturer a nation may be, if it has no means of transportation it cannot distribute its goods or gain the benefits which come from other nations.

Nothing throttles a people's development more than lack of transportation. We have examples of this at our very doors in the Alleghany mountains and on the shores and islets of our Atlantic seaboard. Many of these communities, although composed of the original Anglo-Saxon stock, the first that came to this country, and although they are the purest Americans that we have, not only have made no advance in their cultural state but have retrogressed. Many of these people still speak Elizabethan English and are a prey to the beliefs and superstitions of the Middle Ages. Many other examples of a similar nature, well known to all, can be given. These conditions are entirely due to a lack of transportation. Frequently, I have had forced landings with my airplane in these out of the way communities, among people who were unable to read and write and who did not know who the Governor or Representatives of their state were and who did not know where the nearest postoffice was, although, in one instance, it was only about eight miles off. In this particular community, their only fear was of the United States revenue officers whom they regarded as the only Government that came into contact with them. The evasion of the provisions of law that the revenue officers were detailed to enforce and an occasional feud with a nearby family furnished the causes of the only community organization which they had.

The whole means of transportation on the surface of the ground or water necessarily is confined to places that are easy of access over these elements; in the case of water: deep harbors, indentations along the coasts, and navigable rivers; in the case of land: where it is possible to build roads and railroads. These, in their turn, follow the lines where the grades are the least abrupt and, consequently, are developed along stream lines and across passes in the mountains where the erosion of the water on the heads of the rivers has made the going easier. No condition of this kind confronts aircraft as the air is a common medium all over the world. It is therefore possible to develop transportation to any place desired in this medium. In commercial aviation, however, a positive gain in dollars and cents must be shown over the competing carriers on the land and water. There must be regularity of schedule and the transit of passengers and goods must be safe and not subject to too great a percentage of accidents.

There are two things in which an airplane excels all other carriers: one is its speed and the other is the fact that it is the only instrument of transportation which is capable of delivering its cargo to a terminal station in the air. The latter has been used to great advantage as a means of advertising commodities, such as sky-writing with smoke let out of the airplane in various ways and maneuvering the airplane so as to write letters or words that everybody can see. Another means of advertising is to paint the name of the article on the under surface of the airplane, and still another is to distribute pamphlets or sheets of paper describing the article being advertised. Another use of aviation as a means of delivering something at a terminal station in the air is the photographic camera. The use of this instrument has tremendous possibilities. Not only can it portray the topography of the earth but even the elevations and depressions so that it can be used in surveying the whole country and, in our country, scarcely sixty per cent of the whole area has been adequately mapped. It has taken all of the years since we began to do even that much. A complete photographic survey of the whole country could be accomplished by aerial photography within two years and a half at less than one fourth of the cost required by ground methods. Photographs of agricultural and animal industry areas bring excellent results and often it can be told from the character of vegetation and the color of the ground shown on the photographic negatives, what the best crop to be grown on the land should be, also where irrigation could be used to advantage, water power sites located, and electric power lines installed. In the great fields of sugar cane in the Hawaiian Islands an air reconnaissance is often the only means of telling whether the irrigation is reaching the center of the fields or not. Aerial photographs can portray model farms, feeding places for animals, and shelters.

The Aerial Forest Patrol, inaugurated by the Army Air Service in 1919, saved more for the United States that year than all of the money expended by the Government for aviation.

The civil departments of the Government have uses for aircraft, that, in many instances, do the work much more cheaply than other methods formerly employed and which often are the only means of gaining the information or doing the work desired. The civil departments of the Government have many problems which are yet unsolved. One is the question of making rain. Some time ago the question of attempting to clarify fogs over airdromes led to a very careful study of moisture in the atmosphere. Many interesting experiments have been conducted along that line to show that fogs can be eliminated and that the clouds can be made to deposit their moisture in the form of rain.

These experiments are going forward at this time. Condensation of moisture is brought about by electrified particles of matter. Sand has been used so far, charged with a very high electric potential of an opposite kind from that found in the clouds. This sand is scattered around by aircraft in or over the clouds to produce the effect. The advocates of this method of producing moisture from the clouds have already laid plans for the watering of the arid regions but instead of using sand they will use minute grass seeds which, after they produce the rain, will fall to earth and grow luxurious meadows where thousands of cattle can graze. They also have figured out what the damages would be in case towns or bridges were washed out and how this would be handled. While this is far in the future it is well within the bounds of possibility.

A great deal has been done by the Air Service in the elimination of insect pests, particularly locusts. A remarkable instance of this happened last year in the Philippine Islands. The locusts, which had been feeding high up in the highlands of Mindanao, consumed all the verdure and it was necessary for them to find new pastures to abate their hunger so they descended in swarms on the fertile sugar cane fields at lower altitudes. There were so many that they actually darkened the sky. The appearance of the fields after they had passed was pathetic. Where the tall cane had stood a few hours before, there was nothing left except little short stubs and tattered remnants. The Air Service sent a couple of planes and their crews with what is known as "dusting" equipment, which is a method of throwing out a solution of arsenic and other drugs into the air and down on the ground which, although poisonous to the locusts, does not affect the crops. Immediately millions of these locusts were killed. They were raked up in great heaps. They became so afraid of the airplanes that when they heard them they would fly away and, in this way, many were driven into the sea and drowned. Aviators reported that they could actually be herded and driven in any given direction by the airplanes. Within a few days the pest was stopped. Other instances have occurred where orchards which were being destroyed by insects have been entirely cleansed. So far, no adequate method has been found to eliminate the cotton boll weevil for chemicals that kill the insects also injure the plant or are dangerous to animals and human beings.

The Air Service has been used extensively for medical control in certain places. Siam, for instance, has a very efficient Air Service but is very deficient in land communications so it has inaugurated an air transport system throughout the kingdom. There are a great many poisonous snakes that take their toll of lives annually. In Bangkok, a splendid Pasteur institute for the treatment of snake bites is maintained and persons bitten by snakes or other poisonous reptiles are transferred there by air. All of these lives would be lost if they relied on the ordinary ground or water transportation as it is necessary to treat cases of this kind immediately. Airplane ambulances are being used more and more in all parts of the world.

A distinction has to be made between civil and commercial aviation. Civil aviation is the aviation that is used by the civil departments of the Government. This kind of aviation does not come into competition with the other carriers on the ground and therefore the expense as compared with railroad travel or ocean travel is on a different basis from strictly commercial aviation. Commercial aviation has, however, to compete on a basis of dollars and cents with existing carriers and, in order to be self supporting, has to show a positive economic gain. It is just beginning to be realized by the people of our country what a marvelous thing our Air Mail Service has become. They now demand its expansion all over the country and, after that, will demand its expansion to other parts of the world. A service of this kind is especially valuable to the banker as the time element is the one great cost burden on business transactions. Every added hour in the transit of articles adds to the cost of business in man hours of work, interest carrying charges on commercial paper and, consequently, additional capital assets. As banking deals to a great extent with interest payments and, as interest depends on the time element, the shortening of the time of transportation from one place to another results in a very great saving. The establishment of a central gold fund and the flood of daily telegraphic clearances brought about by the Federal Reserve system, worked such a change in banking methods as to reduce the clearance time by half which means a daily release of fifty per cent of what is called the "float," exceeding five hundred million dollars. It is figured that in the city of New York alone at least one billion dollars in capital is in daily transit in the form of checks.

What has held up commercial aviation, of course, has been the great cost of operation and a lack of knowledge of the articles that can be carried profitably in the air.

Our Post Office Service between New York and San Francisco has proved conclusively that a regular, safe, and continuous service can be established and maintained in the air. Flying can be done both by night and by day on lines of this kind. High or low temperatures and bad weather do not prohibit the establishment of air lines. Airways, consequently, can be established virtually any place in the world. The New York Merchants Association believes that the Air Mail will cause an average saving in time of from twelve to fourteen hours from the cities doing a maximum amount of business with New York. It is said that Cheyenne, Wyoming, estimates a release of three hundred thousand dollars a day in "float" between that region and Chicago and New York, by cutting the time of the present mail deliveries in half. Kansas City has fifty-nine banks in the Tenth Federal Reserve District, with four thousand four hundred and sixteen banks with a business daily clearance of two million four hundred thousand dollars. Nearly fifty per cent of this business is with New York, Minneapolis, Denver, and Dallas. Enumeration of these things can go on all over the country. I have mentioned them only enough to show the great saving that will accrue due to air methods of transportation in financial activities alone. Think of the result of an Air Mail Service between the centers of population in Asia and America which will cut down the time from four or five weeks to from sixty to eighty hours!

The nations of Europe were quick to establish air lines immediately after the World War ceased in 1918. The old war airplanes, transformed and adapted to passenger carrying purposes, were placed on these lines, but, naturally, the expense of operation of this class of airplane was very great. Since then, a constant attempt has been made to develop real commercial aircraft, that is, those in which the cost of upkeep and operation per pound and passenger mile will be greatly reduced. This condition is now in sight and it will not be many years before aircraft are able to compete, in carrying certain classes of goods and passengers, with any other means of transportation on the earth or on the water.

It appears that it costs about as much to carry a pound of freight one mile in an airplane as to carry a ton one mile in a train, or about two thousand to one. The tractive effort necessary to pull an airplane through the air is more than ten times as great per pound of gross weight as by a freight train. The train will coast on a 2% grade whereas an airplane requires about a 20% grade. The unit fuel cost is about ten times as great for the airplane and this proportion may be even greater because a locomotive burns a very low grade fuel and an airplane, very high grade gasoline. The crew of a train, carrying hundreds of tons of freight, is 5 or 6 men; the crew of a weight-carrying airplane averages from 1 to 1½ men for each ton carried.

At the present time airplanes are no faster at distances less than five hundred miles than the existing railroad system, with the time taken to and from the average airdrome, which is usually an hour's drive outside of a city, and the delays incident to embarking and debarking considered. If transportation terminals could be constructed where steamships could dock at wharves alongside of railroads and the whole place roofed over to make an airdrome for airplanes, this feature would be eliminated and the time shortened. Aerial transportation terminals will have to be constantly expanded as the demand grows. A night service for passengers between certain points is a necessity. A good instance is traffic between New York and Chicago. If airplanes flew only in the day time there would be little saving because it requires a night and part of a day at the present time on a railroad train, the daylight portion of the trip being almost as long as the time required by air. If, however, passengers could get in an airplane in New York in the evening and be in Chicago in the morning, ready to do business at nine o'clock, the saving over railroad transportation would be very great. Night traffic is necessary for distances under from five to seven hundred miles, otherwise practical competition is difficult with existing land services. For over water services these distances can be about halved because steamers are slower.

As to carrying passengers, the cost appears to be anywhere from eighteen to seventy-five cents per passenger mile with full loads. Safety of operation along properly administered and installed airways is as great if not greater than for means of transportation on the ground. In military aviation there will always be a certain number of accidents because the military service has to have the fastest pursuit ships, the greatest weight carriers for the bombers, and maximum performances of all sorts which cut down the factors of safety. They have to act in large bodies where the danger of collision is always present. Military aviation is designed to inflict the greatest loss possible against the enemy and the dangers incident to this have to be sustained. In commercial aviation, however, every measure has to be taken for the safety of the passengers and crews so that already very great safety has been attained and the future promises to hold out still more. Most of the accidents to commercial aviation occur during storms or fogs. Instruments now make it possible to maintain safe flight through fogs. The radio telegraph warns of storms so that either landings can be made or the storms avoided if they are too severe.

In speed the airplane excels and will continue to excel, in increasing proportion, all other means of transportation.

The development of strictly commercial types of aircraft will gradually cut the cost of operation and maintenance down. Experimentation along these lines, however, is very costly and the governments have to take the initiative in order to demonstrate how these things can be brought about. In Europe a very heavy system of subsidizing is in vogue. They all operate somewhat along the following lines. If a company desires to go into the airplane transportation business between two localities on the Government airway, it is investigated and found out if the company is a reliable one. Then the Government assists the company by paying about half the price of the airplanes and equipment. This equipment is subject to Government inspection frequently to insure its being kept in good condition. Types of airplanes used so far can practically all be converted to military uses. Pilots and mechanics have to pass Government inspection so as to insure the aircraft being operated by reliable men. Companies get a certain subsidy for the number of pilots and mechanics they maintain. Last of all, they are guaranteed a certain net income per year, usually five per cent. If the profits are below this, the Government makes up the difference. If they are more than this, the companies keep what they make and the Government pays nothing. A system of this kind not only develops commercial aviation but also helps to maintain the airway crews and equipment at only about half of the cost that the Government would have to pay if it maintained them all itself. The underlying motive in these services is military and the commercial part of it is entirely secondary. Great nations, however, seeing the coming of air transportation in the future and knowing its potentialities, are laying plans for monopolizing this means of transportation in the future.

In America, no system for the development of commercial aviation has ever been developed. As only a few small airplane ventures pay, comparatively little has been done along that line as compared with the airplane lines for passenger and freight plying between London, Paris, Brussels, Berlin, and other parts of Europe.

It would seem logical in this country that a corporation should be organized by the Government to do the pioneering in the development of commercial air transportation. Such a corporation could establish commercial airplane lines along the existing postal airways and carry express, freight, and passengers. Accurate cost accounts should be kept and the expenses incident to this kind of traffic and the best equipment essential for commercial flying should be made public so that whenever a civil corporation desired to begin operations on its own hook it could do so with the assurance of what its outlay would be and what the returns might be. While this was being done, a commercial survey of the country, to determine what articles can be carried at a profit through the air, should be made. The Government really is the only agency in our country that could do a thing of this kind as it involves a great deal of expense and investigation. If such a system were adopted, there is no question but that the United States would soon lead in commercial aviation. After lines had been established throughout the country, other routes could be established to South America, Asia and Europe.

An entirely new development along long distance routes will be the study of the air currents. There are trade winds in the upper atmosphere just as there are trade winds near the ground. We know that the atmosphere extends up to fifty-five miles and that, in all probability, aircraft can be made to navigate in most of it. We are not certain what the conditions are at the great altitudes but we do think that we can make greater speeds, in the rarified atmosphere and that, with these greater speeds, if we take advantage of strong air currents that blow continuously, we can cut down our time of transit very greatly from continent to continent. A real meteorological survey of the world at altitudes up to seven or eight miles above the earth should be completed.

The Germans, before the war, did more than any other nation in the development of transportation by air. They had perfected their dirigibles, or Zeppelins as they are called, so that it is said that over two hundred thousand passengers were carried by these ships without accident. The cost per passenger mile is very much less in lighter-than-air ships than it is in airplanes and the limit of cheapness has not yet been attained. Larger airships will be much more cheap to operate, comparatively, than the smaller ones. Passengers can be carried in airships at a rate of about three cents a mile or even less, and the speeds made by them, point to point, will be almost double that of railways. The installation for the use of airships, of course, will cost a great deal more than that for airplanes. Large hangars will have to be constructed and provisions made for handling the airships in heavy winds and during storms. In France, re-enforced concrete structures, one thousand feet long and one hundred and sixty feet clear on the inside, are used. A single dirigible installation might cost ten million dollars which is not excessive for an airport when compared with similar installations for railways and steamships. It is said that the Pennsylvania Railroad station with its terminals in New York, cost over two hundred million dollars; the Washington station with its terminals cost thirty millions; the Lake Shore station of Chicago with its terminals cost sixty millions. Against this, an airport for dirigibles at New York and one at Chicago would cost about twenty million dollars and would house airships sufficient in size and numbers to carry as many passengers as are now carried by the fast trains between New York and Chicago.

It is difficult to convey an idea of the comfort of airship travel. There really is nothing like it. The cabins are large and commodious. One can walk around. There are no severe shocks or jolts such as are experienced on a railroad train and almost no undulations such as are experienced on seacraft. There is no dust, no noise, and any temperature desired can be maintained. The view from the cabin windows gives one a better idea of the country than it is possible to get by other means of transportation. When people understand the safety of this method of transportation it will be very popular.

These airships have the power of remaining in the air longer than any other aircraft and can be constructed to have the greatest cruising radius of any known means of transportation. They can be designed to cross the Pacific just as easily as the British airship R-34 or German ZR-3 crossed the Atlantic. Undoubtedly, if the Germans had not been prevented from developing their airship service as a condition of the treaty of Versailles at the conclusion of the War, they would have airship lines all over the world by this time.

The Zeppelin Company is a remarkable institution. Its capital stock was subscribed by the German people after Count Zeppelin himself had expended his entire fortune in experimentation on these wonderful aircraft. No dividends, of course, are paid out from this stock. Subsidiary companies which make the cloth for the covering of the airships, which make the duralumin for its beams and internal structure, which make its engines, which make its gold-beaters' skins for the bags to contain the gas, which make the gas, and which make the innumerable things which go into the airships' construction, and which return profits, put it back into the general fund of the company. In this way a continuous and lasting system of development was provided entirely independent of Government help. The Zeppelin Company still exists. Many of its subsidiaries are working and making money at present. When the time comes either to renew the Zeppelin service in Germany or to go to other countries and develop, it undoubtedly will be done.

The United States Army Air Service accomplished a feat on December 16, 1924, which will have a profound effect on the use of airships in the future. This was the landing of an airplane on an airship. The airplane flew up and hooked on to an especially constructed gear under the airship while both were flying at a speed of about sixty miles an hour. The airplane then remained for some minutes attached to the airship, stopped its engine, later started it and then took off from the airship and landed. This conclusively shows that articles may be delivered from an airship anywhere over a locality, passengers may be debarked or embarked, fuel may be taken on, and last of all, the airship may be used as an airplane carrier. This fact enhances the usefulness of the great dirigible airship many times.

Another feature of commercial aircraft development is the necessity for uniform rules of the air, examination of the pilots, and examination of the airplanes in a way similar to that which is done for ocean shipping. We have no federal laws governing matters of this kind so that anyone, subject to some local state regulations, can take out aircraft and operate them no matter whether they be safe or dangerous. Each state or community, of course, can prescribe what it sees fit as to the operation of aircraft and, in the future, unless the Federal Government acts, it is quite possible that there will be as many different regulations as there are states which will greatly interfere with aerial navigation. On the other hand, regulations must not result which will crowd out the small operator and interfere with the development, particularly, of light planes. During the last summer I saw a little airplane fly from Dayton to Columbus, Ohio, and back, a distance of one hundred and sixty miles, on two and one half gallons of gasoline, or over sixty miles to the gallon. If strict regulations are adopted, a certain amount of assistance should go with them to insure development, otherwise it will be all restriction and no assistance. On the other hand, intelligent supervision based on a knowledge of air matters should be incorporated into the organization, handling, and enforcing the rules. In England, before the establishment of their department of the air, this regulation was turned over to the British Board of Trade and it is said that one of the first regulations they made was that when two airplanes met each other in the fog they should blow their fog horns! There were other rules almost as ludicrous as this, because the work was being done by men untrained in air matters.

Our airways should be properly organized with distinguishing marks along them so that the aviators can see to insure proper flying by day and by night. Night flying is just as secure and even more easy than daytime flying as the direction can be maintained easily up to about twelve o'clock at night because the cities and small towns make very good beacons, but after that, on account of turning out their lights, it is difficult to keep one's direction without a regular system of night guiding lights. There will be directional radio and wireless systems to guide the airplanes; good weather systems to notify them of storms so that they can shift to the north, south, east, or west, away from them; and instruments for flying and landing in the fog. When a system of this kind has been put into effect, passengers and light weight articles would be able to go from New York to San Francisco safely and on regular schedule in from twenty to thirty hours. A service of that kind could be inaugurated at the present time on a basis of sixty cents per passenger mile and, if full loads were carried, it could probably be operated on a basis of eighteen cents per passenger mile. This would make the cost for such a trip about four hundred and fifty dollars but would be four times as quick as the railroad trip and much more comfortable.

An all-land airway can be established to South America and take passengers from New York to the Argentine Republic in from fifty to sixty hours and also a practically all-land route from New York to Peking, China, by way of Canada, Alaska, and Siberia, in from sixty to seventy hours. Airship traffic, that is, the large dirigibles carrying much heavier cargoes both in freight or in passengers than airplanes are capable of handling, could cover the same distance in from two to three times as many hours, that is: they would be two or three times slower than the airplanes. Both of these modes of transportation are from four to ten times as fast as existing means of transportation on the ground or on the water. They are not confined by either one of these elements and can go anywhere that there is air. Their development is only a matter of time and this time should be shortened by the intelligent direction of the Government because the initial outlay and the experience required, necessitate too great an outlay for any civil corporation to take up in their entirety. Not only will every part of the world be reached but the world itself will be made correspondingly smaller because distance will be measured in hours and not in miles. The substantial and continual development of air power should be based on a sound commercial aviation. America is in a better position to develop commercial aeronautics than any other nation in the world.

## V

## HOW SHOULD WE ORGANIZE OUR NATIONAL AIR POWER? MAKE IT A MAIN FORCE OR STILL AN APPENDAGE?

"Where there is no vision the people perish." This old Biblical quotation is more applicable to the development of aviation and air power than to any other undertaking.

We are at the turning of the ways in the development of our air power and the people, who are the judges of what should be done, should weigh the evidence on the subject carefully. In order to be successful in anything, it is necessary to concentrate one's mind, one's time and one's money on it in such a way as to get the greatest good with the least effort. In doing this with aviation, vision is a most important matter because its great possibilities lie ahead and not behind us. At this juncture, the United States is faced with the alternative of progressing in its aeronautical organization and consolidating its air activities under one responsible head, or going on with its effort split up between other services that have a major function apart from aeronautics.

Aviation is very different from either armies or navies in its economic aspect. Every military airplane can be used in time of peace for some useful undertaking not necessarily connected with war. Every pilot employed in civil aviation can be used in case of war and is ninety per cent efficient at least in time of peace. Every mechanic used in civil aviation is one hundred per cent efficient in time of war. In time of peace, the bulk of the effort and thought of a nation in an aeronautical way may be applied to civil and commercial development of aeronautics and this same effort and thought can be shifted at once to military purposes. There is no reason, for instance, why the air forces in time of peace should not be employed in mapping the country, patrolling the forests to prevent forest fires, carrying the mail, eliminating insect pests from cotton, fruit trees and other vegetation, and in making an aeronautical commercial transportation survey of the country to determine what can be carried economically and at a profit through the air instead of on boats, railroads, and by automobiles, and in working out new commercial air routes throughout the world. The Government, for instance, in time of peace should maintain only a small percentage of its total aerial strength on strictly military duty; the rest could be used on civil work for the greater part of the time and assembled for a month or so each year to perform maneuvers and military training.

The great countries of the world are using their vision and are straining every effort to establish their aeronautical position so that the future will not see them hopelessly distanced by their rivals.

So far as national defense is concerned, they have carefully studied the whole problem as affected by aviation, so that they will get a maximum benefit from each dollar of money expended and from each man hour of work put in. From a military standpoint, the airmen have to study the effect that air power has on navies and what their future will be. They know that within the radius 'of air power's activities, it can completely destroy any surface vessels or war ships. They know that in the last war, surface ships, battleships, cruisers and other seacraft, took comparatively little active part except as transportation and patrol vessels. No battleship sunk another battleship and of the hundred and thirty-four warships sunk or destroyed during the war, the submarines sunk sixty-two British warships and eight large French and Italian ships. No American battleship saw any fighting in the last war, not even those in European waters. Aircraft have great difficulty in attacking and destroying submarines at sea. They are very hard to detect, dive with great rapidity and are very difficult to see under water. The effect of air power on submarines is probably less than on any other target, whether on water or land. The best offense against them is to destroy their bases and fuel stations. It is necessary to consult the best available information about them as they will be the future means of operating on the seas. Existing records show that submarines sunk, either by torpedoes or mines, the battleship, Audacious; they sunk the cruiser, Hampshire, with Lord Kitchener on board; they sunk the cruisers, Cressy, Aboukir and Hogue in a few minutes. From that time on, the British battleships were either tied up in their ports behind torpedo nets and screens of destroyers and submarines or they were ziz-zagging their course at great speed for a few hours on the high seas. It is stated that the submarines sunk two battleships in the Dardanelles and drove the Allied fleets into Mudros harbor. Men even landed from a submarine in Turkish territory and blew up a bridge by setting an explosive charge off on it.

A modern battleship, according to the old system of naval thought, may cost somewhere between fifty and seventy million dollars; it may require, on an average, one cruiser costing between twenty and thirty million dollars, four destroyers costing three or four million dollars each, four submarines, a certain amount of air power to protect it, and, in addition to this, great stores for maintaining the personnel of more than a thousand men and dock yards and supply facilities to keep them up. So that every time that a battleship is built, the nation constructing it is binding itself to about one hundred million dollars or more of expenditure and a certain amount per year to keep it up. Battleships have required heretofore complete replacement every few years to prevent their becoming obsolete.

As battleships and surface craft are helpless against aircraft unless they themselves are protected by air power and, as their influence on the destruction of sea-going trade is secondary to that of the submarines, nations are gradually abandoning battleship construction. Three are keeping it up: England, Japan and the United States.

England is entirely dependent for existence on her sea-borne trade; Japan, also, is dependent almost entirely on her sea-borne trade. Where England and Japan would have to protect their commerce in the Seven Seas or starve, America could entirely dispense with her sea-going trade if she had to, and continue to exist and defend herself. Where, therefore, a nation might have to expend a tremendous amount of effort and treasure on the maintenance of its sea-borne trade at great distances from home, it would be better for one not so dependent on sea-borne trade to put its national defense money and effort into active offensive equipment designed directly to defeat the enemy instead of dissipating its power in an indecisive theater.

The airman looks at the development of a country's military effort somewhat as follows. National defense consists roughly of four phases: First, the maintenance of domestic tranquillity in the country itself so that the preparation of active fighting material can go on unhindered. An army on the ground to insure tranquillity and an air force in the air to prevent hostile air raids can take care of this. Second, the protection of the coasts and frontiers. An air force can do this and fight any hostile aircraft or destroy hostile warships while its home country is policed and protected on the ground by a land force. Third, the control of sea communications. This can be done by aircraft within their radius of action and otherwise by submarines. Surface craft have a secondary value for this. Fourth, the prosecution of offensive war across or beyond the seas. This may be carried out primarily under the protection of air power, assisted by submarines and an army. A succession of land bases held by land troops must be occupied and the enemy must be attacked directly through the air. Floating bases or aircraft carriers cannot compete with aircraft acting from land bases. So that, in future, surface transports escorted by war vessels such as carried the American troops to Europe cannot exist in the face of a superior air force. Only when complete dominion of the air has been established can a war of invasion across the seas be prosecuted under present conditions. Air power, therefore, has to be employed as a major instrument of war, no matter whether a land force or a sea force is acting on the surface of the earth.

Submarines have proved themselves to be the great destroyers of commerce. Existing records show that during the war the Germans maintained only about thirty submarines at sea. They started the war with a total of about forty submarines, counting all sizes. That was a small number but they had a good start in their design and development work. As they increased the numbers they improved their characteristics, and the Allies were occasioned several surprises. The radius of action and the periods that German submarines kept the sea, even early in the war, were not previously believed possible. The guns which they installed on deck like any surface craft and fought in fairly rough seas, constituted a novel idea. Mining from submarines was entirely a surprise and a very disagreeable one. The Germans could do little or nothing on the British and French coasts with surface vessels but their submarine mine layers, of which the British or French had no inkling until they began operating, could reach any coasts of the British Isles or France and were at it for years. They not only mined a great many ships but forced their enemies into employing an enormous force of mine sweepers. Large areas off the British and French coasts are mining waters, necessarily used by the extensive shipping and had to be swept daily. Many sweeping details did not find a mine for months but since a mine layer was likely to plant a few of her eggs any night, the sweeping had to continue.

The Germans built or had building when the war ended about four hundred and twenty submarines. They lost, either by the Allies' action or accident while operating, about one hundred and forty-six, and they destroyed between fifteen and twenty in Belgian and Austrian ports before signing the Armistice or while en route to be surrendered. They surrendered about one hundred and seventy, many of which were not in operating condition, either not completed or damaged; and, at the end of the war were building about sixty others. There was also a large additional program started by Admiral Scheer soon after he was put in charge of their entire navy near the end of the war. The Austrians had a small submarine force which accomplished little; they lost eight boats during the war.

The Germans built a total of about four hundred and thirty submarines, of which about one hundred and forty-six were sunk, as follows:

| | |
|---|---|
| By anchored mines, including those in nets...... | 42 |
| By depth charges, from all classes of vessels.... | 35 |
| By gun fire, including those on decoy ships ...... | 24 |
| By torpedoes of submarines ................. | 20 |
| By ramming, including all instances ........... | 18 |
| By air attack ............................ | 7 |
| | 146 |

In the past, the first cost of all classes of surface ships and of submarines had been approximately the same per ton. Submarines wear fully as well as capital ships and have much longer life than light, fast vessels. Their maintenance, fuel, personnel and other running costs are much lower than anything else—whether reckoned per ton or for potential war value.

The submarine's defensive power does not get out of date; the oldest boats are defensively as strong as the newest ones. Their defense depends neither upon speed nor their offensive power or that of other vessels. This unique defense by simple concealment is inherent in the type and is the quality which makes the submarine able to act unsupported and to play a lone hand. No other ship can, unless with long sustained speed sufficient to get away from everything that is stronger.

It is no doubt believed by many that hydrophones or kindred devices have entirely compromised the submarine's defense. These devices are as extensively used in submarines as in anything else. They are at their best in a submerged submarine and such use is most favorable for estimating their value. While diving, they do hear and locate each other at some distance if the conditions are favorable; otherwise they don't. When it comes to dependence upon hearing an extremely faint noise through hydrophones amidst the loud and numerous noises of a formation of surface ships running at usual speed, it's quite another matter.

It is true that some German submarines were followed and destroyed by use of hydrophones. But the average was not high and it is said that the captain of any well-conditioned submarine will bet you to his last dollar, and give you big odds, that he can get away from any hydrophone pursuit which can now be organized. Possibly the listening devices will improve but no advance since the Armistice nor any projected device threatens the submarine very much. It is to be remembered that since the submarine is the ideal listening vessel, its use of those devices is a great aid in outmaneuvering anti-submarine craft.

The fuel radius of submarines is higher than in any other class of vessels. Even small ones run long distances and a large one could make a non-stop run around the earth. Their Diesel engines are more than twice as economical as any steam plant. Their fuel and ballast tanks hold large stocks of fuel which are carried without sacrifice other than sluggishness in a sea-way and some reduction in speed. Since their defense is independent of speed, it is entirely safe to load them down thus with fuel.

By sea-endurance is meant habitability and capacity

for consumable supplies, including ammunition. Long periods of living in a submarine are none too pleasant but have been and can be endured. In fact, if a boat is properly designed, particularly with a view to minimum requirements in number of personnel, it is not so bad. It is possible to work out the capacity for supplies without difficulty and it can be safely said that if properly designed, a sea-endurance can be obtained which balances the fuel radius. These factors give a vessel which can keep the sea for long periods, cover long distances and operate unsupported.

Any valid criticism of war values seems to question the submarine's powers of offense against surface war vessels. At present development we have seen that they were powerful enough to lay mines secretly and where they liked; their guns, which during the war were only patched on, did effective work; while their torpedoes, husbanded for safe shots only, caused heavy losses in men-of-war. Aside from the direct results were the restrictions that their menace placed upon the Allies. It was a nuisance to have to zig-zag, etc. Escorting vessels used up a lot of our resources and energy; for instance, destroyers had to be specialized in armament and tactics at the expense of their value in the missions for which they were built. What the German submarines did against Allied men-of-war, alone, amply repaid them for their total effort.

The direct offensive power of the submarine is still developing. A British boat carries a 12-inch gun which operates successfully. It is easy to mount powerful guns up to 8-inch or more and to provide ammunition supply and fire control. Guns on submarines, probably, are only auxiliary weapons and under-water attack is still their main job. There are also said to be weapons other than torpedoes for attacking under water. These new weapons are mainly in the paper stage only; if successful, they will surely destroy any ship built or building. It seems that a vast improvement in offensive power—certainly all that needs to be added to present-day submarines—is practicable.

Submarine officers think our next national emergency will find them fighting on our most advanced front from the day hostilities begin. It is conceivable and probable that there will be a long period of hostilities before any surface fleets come into action. The weaker surface fleet would certainly retreat to the protection of its air power in the radius of aircraft action of its own coast. The superior fleet menaced by submarines and long distance aircraft could not long exist on the high seas and would be of little service there under such conditions. A fleet action in the old sense may never occur again. Undoubtedly, submarines will be developed into aircraft carriers in addition to their other uses.

A modern organization of a country's military power, therefore, indicates that aircraft will be used over both land and sea for combating hostile air forces, demolishing ships on the sea and important targets on the land. Submarines will be used in and on the seas for controlling sea lanes of communication and assisting air power. Armies will be used on land for insuring domestic tranquillity, holding operating bases for aircraft and seacraft and, in a last analysis, together with air power against hostile armies. What might be termed "battleship sea power" is fading away. Only a few nations still maintain it. If an attempt is made to use it in future, it will be so menaced by aircraft from above and submarines from underneath that it will be much more of a military liability than an asset. To this extent has the advent of air power and the use of submarines wrought a change in methods of war.

In the future, therefore, surface navies based on battleships, cannot be the arbiters of the communications over the ocean.

The tremendous cost of these craft and their upkeep will be applied to more efficient and more modern methods of defense. Fighting airplanes can be built in production with their engines for from fifteen to seventy-five thousand dollars, or an average of about twenty-five thousand. Therefore, so far as construction is concerned, at the price of a battleship and its accessories, that is: one hundred million dollars, an average of four thousand airplanes can be built for the price of a battleship.

The United States is now allowed a fleet of eighteen battleships. On this basis, seventy-two thousand airplanes could be built. In any national emergency that we can visualize, the country certainly does not need over three or four thousand airplanes at the decisive point; these can be built and maintained for a relatively small proportional cost and still have great use in civil and commercial aviation in peace time. The cost of the battleships and their accessories is not all. The Navy Yards cost tremendous amounts. In the United States alone there are some twenty of these, whose value aggregates one billion three hundred million dollars. The cost of upkeep and depreciation of these amounts to a vast annual sum. Into many of these Navy Yards a wounded battleship drawing forty feet of water cannot go as there is not enough water on the sills of the dry docks.

If the defense of the coast is intrusted to aircraft and the Navy's coast defense functions are modified, many of these stations can be dispensed with.

In order to carry on offensive operations, a surface navy has to have tremendous naval stations and bases, thousands of miles away from their own country in some cases. These take the forms of dry docks, fuel stations, oil and ammunition depots and shops which cost millions of dollars and are quite vulnerable to air attack. The amount of money and effort put into these might be applied to better use for aircraft and submarines.

So far as land forces are concerned, airplanes will reduce the cost of coast fortifications. As they are able to attack seacraft at long distances from the coast, they not only will keep surface seacraft entirely away from cannon range of the coast but they will eliminate the necessity for many of the great seacoast cannon. Every time a large seacoast cannon is installed on its concrete foundation, it costs half a million dollars. In the ten years preceding 1920, the United States expended about one billion eight hundred million dollars on coast defenses of different kinds. The present land system of fortifications has changed little from the system employed during the Revolutionary War, which was to have all estuaries, ports, or harbor entrances garnished with cannon so as to keep away all hostile surface ships. Part of the money and effort saved from some of these expenditures could be put into more mobile cannon to be used with an active army, or into aircraft to keep the enemy away from the coast and frontiers. With a well organized air force, it is hard to visualize how an enemy could gain a footing on land in a country such as the United States.

Constant development must be kept up in civil and commercial aviation as well as in military aviation, and again accurate vision is required to see what will take place several years hence. The older services such as the Army and Navy to which aviation was attached at first, were entirely incapable of visualizing aviation's progress, particularly in its civil and commercial application which must work hand in hand with its military use.

Another very important consideration is the budget. So long as the budget for the development of aircraft is prepared by the Army, Navy or other agency of the Government, aviation will be considered as an auxiliary and the requisite amount of money, as compared with the older services, will be subject to the final decision of personnel whose main duty is not aviation. This has resulted in an incomplete, inefficient, and ultimately expensive system of appropriating money.

Of equal importance is the question of the personnel,

of the people who have to act and actually fly in aviation. We now have an air-going personnel as distinguished from a sea-going and land-going class. In military aviation, in time of peace, the Air Service in the United States loses nearly half of the total number of deaths in the Army per year; in time of war, the number of casualties among the flying officers is proportionally great. We therefore need an entirely different system of training, education, reserves, and replacements, from that of the other services.

As important as anything else is the placing of one man in charge of aviation who can be held directly responsible for the aeronautical development of the whole country and, next, an air representative on councils of national defense, who has co-equal power with that of the representatives of the Army and of the Navy. Not only does this give proper weight to aeronautics, both in peace and in war, but the Army and Navy have always, and will always, deadlock under certain issues where they have equal representation. The introduction of a third service would tend to break this. Eventually, all military power of the Government should be concentrated under a single department which would have control over all national defense, no matter whether it be on land, on the sea, or in the air. In this way, overhead might be cut down, definite and complete missions assigned to air, land and water forces, and a more thorough understanding of the nation's needs would result.

When the great nations considered these things and many others, they gradually began changing their governmental organization to keep step with the progress of the time.

England created a separate air ministry co-equal with the Army and Navy. This department handles all aviation matters: the central air force, the aviation assigned to the Army and Navy, civil aviation and commercial aviation. It maintains its airways, weather services, radio control stations, and subsidizes its passenger and cargo planes. One air man has charge of all of England's air defenses. An air man sits in the Council of Imperial Defense. He has an equal voice with the representatives from the Army and the Navy.

It is probable that both the Army and Navy will be under the Air Commander's orders for the defense of the islands; certainly in the beginning of a campaign, as the paramount interest of the services in the event of an attack on the British Isles resides in the air. Later, if a campaign or war develops so that the Army or Navy has the paramount rôle, the supreme command would pass to the one which is most interested. Wherever the air force can administer an occupied territory more economically and better than the Army, such territory is turned over to them. For several years, now, England's air force has had control and the complete administration of Mesopotamia.

The British air force is composed of men who have complete confidence in the future of aviation and who can visualize what is going to happen and what aviation can do and should do instead of what it cannot do and should not do as the armies and navies were inclined to do when aviation was under them. The British air force has given excellent service to the Army and has greatly improved the aeronautical equipment for the British Navy. Its influence on the design of the latest capital ships of the British service has been marked. These are really great armored airplane carriers; their guns and their planes can probably destroy any other surface ships that now exist. It appears that they have made all the present battleships as obsolete as when the original Dreadnaught appeared and made all the others obsolete. It is a matter of discussion now whether it would not be better to wipe out every battleship and begin all over again, putting one's faith in entirely new developments which can compete with and be more efficient than these new carriers on the high seas.

France has preened her wings also. She has abandoned the construction of battleships, has constructed submarines and has developed the greatest air force in the world. She has a separate department of aviation. The actual fighting part of the air force still remains under the Army, but the development of aviation is now intrusted to a department. It is better than the old arrangement but not as good as England's organization and there is strong agitation for an air ministry to control all air matters.

Italy is organizing a separate department of aeronautics similar to that of England.

Germany had a separate air service in 1916.

Denmark is abandoning her army and navy and relies for protection on her air force and police.

Sweden has an air ministry and is concentrating her power on the development of the air.

Japan is diving into the aviation pool as deeply as possible. She still has an inefficient organization but is consolidating her aviation activities more and more.

Russia is developing her air power and has a single department of national defense.

America still hesitates to consolidate her aeronautical activities but the question is becoming more important every day and the more it is investigated, the more apparent is its necessity. The only American mission consisting of representatives from civil life, the Army, the Navy, the Council of National Defense, leaders in the aeronautical industry and headed by the Assistant Secretary of War, that ever made a careful study of this matter, reported as follows to the Secretary of War on July 19, 1919.

To the Secretary of War.

Sir:

In accordance with your instructions, the American Aviation Mission visited France, Italy and England. It was able to confer with various ministers of these Governments, ranking Army and Navy commanders, and the foremost aircraft manufacturers.

A thorough study and investigation were made by your Mission of all forms of organization, production and development. As a result of these studies your Mission desires to emphasize the universal opinion of its members that immediate action is necessary to safeguard the air interests of the United States, to preserve for the Government some benefit of the great aviation expenditures made during the period of the war, and to prevent a vitally necessary industry from entirely disappearing. Ninety per cent of the industry created during the war has been liquidated. Unless some definite policy is adopted by the Government, it is inevitable that the remaining ten per cent will also disappear.

In placing this matter before you the subject falls into three important heads:

1. General organization
2. Development, commercial
3. Development, technical.

### I. GENERAL ORGANIZATION

The findings of the American Aviation Mission and its recommendations are submitted after a careful review of the situation in the allied countries mentioned, but always keeping in mind the situation in the United States. Under the above sub-heads the results of these investigations are presented to you, which, in the opinion of the Mission, demand the most earnest and immediate consideration along the broadest lines, with a view to establishing some fixed policy which will save the aircraft situation in the United States and give the United States an equal place with the great powers of Europe in this great new commercial development.

The American Aviation Mission therefore recommends the concentration of the air activities of the United States, military, naval and civilian, within the direction of a single Government agency created for the purpose, co-equal in importance with the Departments of War, Navy and of Commerce, to be called in this report, for the purposes of identification, the National Air Service.

In making the above recommendations, the following views and data of the Mission are presented:

Visits were made by the Mission to England, France,

Italy and conferences have been held with those largely responsible for the successful prosecution of the war and especially with those men most experienced in the aerial development within those countries. Among others, interviews have been had with:

FRANCE: Maréchal Foch, Commandant-en-Chef des Armées Alliées; André Tardieu, Ministre des Affaires Franco-Américaines; Général M. Duval, Chef de Service de l'Aeronautique; Jacques Dumesnil, Deputé, formerly Sous-Secrétaire de l'Aeronautique; M. Loucheur, Président du Conseil de Guerre, now Minister of Reconstruction; Daniel Vincent, Deputé, formerly Sous-Secrétaire de l'Aviation; Gaston Minier, Deputé, Chef du Comité Aeronautique au Senat; and Major d'Aiguillon, of the Commission Interministerielle de l'Aviation Civile.

ENGLAND: Honorable Winston Churchill, M. P., Secretary of State for War and Secretary of State for Air; Field Marshal Sir Douglas Haig, Commander-in-Chief of the British Army; Admiral Sir David Beatty, R. N., Admiral of the Fleet; Major General Right Hon. J. E. B. Seely, Under Secretary of State for Air; Major General Hugh M. Trenchard, Chief of Air Staff, Royal Air Force; Major General E. L. Ellington, Director General, Supply and Research, Royal Air Force; Major General Sir Frederick H. Sykes, Controller General Civil Aviation, Royal Air Force; Sir W. A. Robinson, Secretary, Air Ministry; and Major General Sir W. S. Brancker, Royal Air Force, now with the Aircraft Manufacturing Co., Ltd.

ITALY: G. Grassi, Chief of the Italian Aviation in Paris; Colonel Guidoni, Italian Foreign Aeronautical Mission; Admiral Orsini, Chief of Italian Naval Aviation; Colonel Crocco, Chief of the Technical Bureau; and Signor Conti, Secretary of State for Aviation.

In all countries visited, and in the minds of all persons met in conference, appears an extraordinary similarity in condition and in conclusions drawn from the experiences of the five difficult years of mistake and achievement in the prosecution of the war. Perhaps no stronger or more simple presentation of the regard in which the future of aviation is held in allied countries can be given than by quotation from two letters of M. Clemenceau, copies of which were obtained in France. The first is addressed to the President of the United States, urging upon him the immediate consideration of matters aeronautical and in connection with the Peace Conference. The second is addressed to the President of the Republic of France, submitting the draft of a decree creating a separate department of Aeronautics placed transitorily under the Ministry of War—an intermediate step possible without legislation and looking to the early creation of an Independent Ministry of the Air.

Those interested in the future of the country, not only from a national defense standpoint but from a civil, commercial and economic one as well, should study this matter carefully, because air power has not only come to stay but is, and will be, a dominating factor in the world's development.

# VI

# THE EFFECT OF AIR POWER ON THE MODIFICATION AND LIMITATION OF INTERNATIONAL ARMAMENTS

THE rapidly increasing efficiency of the airplane and the submarine gives us the opportunity to move towards a new limitation of armaments.

Both of these implements of national defense are essentially defensive in their nature as distinguished from offensive military arrangements designed for aggression across and beyond the seas.

They will cause new economies in national expenditures. For example, more than 1000 airplanes can be built and maintained for the outlay required for a single battleship.

Airplanes have a great application in time of peace in useful civil and commercial pursuits. The same airplanes can do this work that are suitable for duty in war and for national defense. In fact all aircraft developments, the factories that make them, the airways that are established for civil aviation and the civilian pilots and crews, are distinct military assets, and can bring in a return in time of peace, thereby reducing the national expenditure necessary in their maintenance if they were kept solely and exclusively for war.

In the case of the submarine, the cost of construction, ton for ton, is about equal to that of other vessels. Their size, however, ranges from 1500 tons to 2500, or less than one-tenth of that of battleships and cruisers. Their efficiency in offensive operations on the sea against any other vessel, either on the surface of the water or below its surface, is constantly increasing. The best defense against submarines is other submarines.

In case of war at the present day, submarines would be the greatest controlling element on the water of the sea lanes of communication, while aircraft above the water would control communication within their radius of operations. The theory of battleship sea power is becoming obsolescent and should be discarded at an early date.

Leaving out of consideration civil, commercial and economic methods of carrying on competitions between the nations and viewing the military side of international disputes alone, it is entirely practical to make up-to-date appraisal of what modern military forces consist of and what they are actually worth. As long as it was a question of land power and sea power, that is, armies and navies, the matter was well understood because they had been continued and applied for centuries. The variation in their use has been very small and has consisted almost entirely in improving instruments and equipment but not methods.

The action of armies and navies on one plane or dimension—that is, on the surface of the ground or the surface of the water—is slow in execution as compared to operations in the air. They also require tremendous and expensive organizations for their maintenance and upkeep on account of the great number of men and the vast amount of equipment which they need. The advent of air power has completely changed this. The relations of armies to navies and navies to armies are now very different from what they were, while both bear an entirely new relation to air power from that which they formerly bore to each other. Even if hostile armies and navies come into contact with each other, they are helpless now unless they can obtain and hold military supremacy in the air. Air power holds out the hope to the nations that, in the future, air battles taking place miles away from the frontiers will be so decisive and of such far-reaching effect that the nation losing them will be willing to capitulate without resorting to a further contest on land or water on account of the degree of destruction which would be sustained by the country subjected to unrestricted air attack.

Air power carries out its military missions, competes in battles in the air and attacks ground and water establishments without participation in its conflicts by either armies or navies. A striking thing about air power, also, is that in time of peace military air power may be employed for useful purposes such as mapping the country, carrying the mails, patrolling against forest fires, aids to agriculture in eliminating insect pests such as locusts and boll weevils, farm surveying, life saving, and an infinite number of other things. No other military formations which the countries possess have such an economic application. From a military standpoint, air power is the only agency that is able to defend the country from hostile air attack and is the principal defense against hostile sea attack along a coast. No defense from the ground is capable of stopping air raids over a country. Along the coast line air power is a positive and absolute defense against any hostile surface ships, as it can sink or destroy any vessels that ever have been built or that can ever be constructed. This feature of the application of air power will constantly increase in its relative ability rather than decrease. Air power, however, has very little effect on submarines and undersea boats. Consequently, naval operations in the future will be conducted by submarines to an increased extent.

The battleship is so expensive and difficult to maintain and so vulnerable to aircraft and submarines that it will be eliminated eventually. Sea power as expressed in battleships is almost a thing of the past. Battleship fleets can no longer control the sea lanes of communication. This attribute has passed to aircraft and undersea boats. The tremendous power of submarines is just beginning to be understood by the people, as the facts and figures relative to their use were concealed largely for political purposes during and even since the War. During the War the Germans maintained at sea only about thirty submarines. These thirty submarines sank eleven million tons of allied shipping. They sank six million six hundred ninety-two thousand six hundred and forty-two tons of British shipping and in nine months of the War, during 1918, they sank one million six hundred sixty-eight thousand nine hundred and seventy-two tons of British shipping. Forty percent of all British shipping was destroyed by these few submarines, which reduced the British Isles to the verge of starvation. The Germans never employed over ten thousand men in the handling of their submarine forces, whereas the allies employed over a million men in counter submarine operations and spent hundreds of millions of dollars for equipment. All of these measures were comparatively ineffective.

Only a total of 146 German submarines were destroyed by the allies during the whole war, out of about 420 built, and out of 134 allied surface warships sank during the war the submarines sank 62. They sank the British battleship Audacious, the British cruiser Hampshire with Lord Kitchener on board, the American cruiser San Diego off Fire Island near New York, and nearly sank the American battleship Minnesota off the mouth of the Delaware River. During the war no battleship sank another battleship and no American battleship entered into combat with the enemy. This was just in the beginning of submarine development and the first application of submarine operations during any war. They are now well organized and have methods of undersea attack by torpedoes, by floating mines and by gun fire on the surface of the water far superior to anything used during the War. They are able to carry guns of any size. Their radius of action is sufficient for them to go round the earth on one charge of fuel. The nature of their warfare is entirely different from that of surface craft. Their great attribute is concealment below the water and this is also their principal defense. They are able to submerge so quickly that aircraft or surface sea craft have little effect on them. The best defense against submarines is other submarines.

When war between two powers separated by oceans occurs in the future all sea lanes leading to the hostile power will be planted with mines by submarines. The surface of all oceans will be districted and organized in geographic squares. To each square submarines will be assigned and a constant patrol will be kept, no matter whether there be warships or commercial ships of the opposing side crossing the seas. The submarines' power is sufficient to destroy any surface vessels including battleships or other war craft. Nations, therefore, will put their money and effort into submarines instead of battleship fleets because the submarine operations are more economical in financial expenditure, in numbers of men employed, and in effectiveness. They, however, are largely a defensive element. They cannot transport troops across the seas as can surface vessels. At present surface craft only may be used for this purpose and also to act as airplane carriers for transporting planes to hostile shores and launching them. As airplane carrying vessels are of no use against hostile air forces with bases on shore, and as they can only be of use against other vessels or hostile fleets that are on the surface of the water, and as these fleets will be supplanted by submarines, there is little use for the retention of airplane carriers in the general scheme of armaments.

Air power has caused a decided change in applying armed forces both defensive and offensive.

In the case of an insular power or continental power, separated by oceans and seas from a possible enemy, it is difficult to see under modern conditions how such a country could be invaded by hostile land forces, or how a surface navy could compromise the country's independence. This is so because air forces can destroy any surface ships that approach the coast of a nation. Submarines also can plant the coast with mines so as to make it practically an impossibility for hostile shipping to land armed forces.

Future invasions into the heart of an enemy country, therefore, will be made by air craft.

In the former conception of national defense, the principle was held that to invade a nation the piercing of its lines of resistance was necessary. If the nation lay across the sea its line of battleships had to be pierced, destroyed and overcome to gain access to the shores. Again, the line of the armies had to be pierced to gain access to the interior. This condition no longer exists in its entirety and is decreasing relatively every day. Air craft do not need to pierce the line of either navies or armies. They can fly straight over them to the heart of a country and gain success in war.

To gain a lasting victory in war, the hostile nation's power to make war must be destroyed—this means the manufactories, the means of communication, the food products, even the farms, the fuel and oil and the places where people live and carry on their daily lives. Not only must these things be rendered incapable of supplying armed forces but the people's desire to renew the combat at a later date must be discouraged.

Aircraft operating in the heart of an enemy's country will accomplish this object in an incredibly short space of time, once the control of the air has been obtained and the months and even years of contest of ground armies with a loss of millions of lives will be eliminated in the future.

Much, if not most, of the present equipment for making war is obsolete and useless and can be replaced by much more economical and useful arrangements and agencies. Nations nearly always go into an armed contest with the equipment and methods of a former war. Victory always comes to that country which has made a proper estimate of the equipment and methods that can be used in modern ways. Air power has introduced new considerations which should be weighed carefully in estimating their effect on the possibility of the limitation of armaments. When the actual value or uselessness of the various branches of national armaments is realized and the conditions affecting their use are fully understood by the people, it will be an entirely practical proposition to limit them. People hesitate to give up arrangements for insuring their own safety.

In countries that are susceptible to constant altercations with their neighbors every man, every woman and almost every child knows what the underlying principles of their national defense are, and they freely and unhesitatingly give their services, their money and their time towards protecting their homes, their institutions and their governments because they have been taught the necessity of this by repeated examples during practically every generation of their existence. They cannot maintain their lives in peace and tranquillity without protection. In other countries far removed from possible aggression, the people have turned for provision for their security to professional organizations created for that purpose and have ceased to interest themselves personally in these matters because they consider the possibility of war so remote. These rely entirely on the advice of professional soldiers and sailors. Usually nations of this kind are the ones that suffer more terribly in casualties, financial outlay and disaster when war takes place, because the conservatism of these permanent military services always tend to perpetuate their existing systems and institutions and resist changes and innovations.

They always fear to change, to do away with or eliminate anything which has long been a part of their organization or system. Unless the public and the legislatures periodically inspect and overhaul the professional organizations maintained for national security, increased expenditures, adherence to obsolete and useless principles of defense, and an inexact knowledge of military conditions are always the result. Contests between nations consist of many things beside actual armed conflicts. They start with competitions of various kinds, usually commercial, and are carried on in different ways by the nations involved depending upon their means of transportation and communication, their wealth and their diplomatic aptitude.

Armed forces usually are called in only as a last resort and when all other means of coercion have failed. The maintenance of armed forces by a nation may be regarded as a direct symptom of the country's punitive psychology. A very accurate estimate may be made as to how and in what manner the policies of a state are to be enforced by the size of its organized armament and the distribution of its forces.

Within the last generation great changes have taken place in the internal organization of states and in their relations to each other. A constantly increasing influence in the policies of a State is being wielded by the common people as distinguished from the ruling class or an aristocracy, as was formerly the case. This condition has been brought about through two principal agencies. First, the general education of everyone which makes it possible for most of the people to read, write and transmit their ideas from one to another. This encourages freedom of speech and a discussion by the people of the nation's policies, both national and international. Second, the development of electrical communication which puts every part of the world in constant touch with every other part and makes possible an instantaneous distribution of the thoughts and ideas of individuals to an increasingly literate people.

Secret diplomacy in its old application is difficult

to carry on today because it is next to impossible to conceal warlike preparations.

The nations understand each other better than has ever been the case before. The factors which have contributed decidedly to the understanding of one people by another have been the constantly increasing rapidity of transportation both on the land and over the water, the railroad and the steamship. The facility with which voyages now can be made around the world is not only increasing commerce between the nations, but the travel of individuals between the hemispheres. There is no part of the civilized world that cannot be reached at present in a fraction of the time that was required fifty or a hundred years ago. Within the last decade the advent of air transportation has added a decidedly new element in the relations of nations to each other.

The older means of transportation, vessels and railways, have followed the parallels of latitude, through the Temperate Zone. The new means of transportation through the air will follow the meridians, that is, the shortest routes, straight across the poles in the north and south—the most direct lines from place to place. To illustrate this, we may take for example a line of communication from New York to Peking, China. At present the route by sea and land goes across the United States to the Pacific Coast, from there via the Hawaiian Islands, Japan, and the littoral of Asia to Peking. It takes from four to five weeks for the voyage along that line in railways and steamships. By going directly across the top of the earth from New York via Lake Athabasca, Canada; Nome, Alaska; Khabarovsk on the Amur River in Siberia, and thence to Peking, the trip may be made by air in from sixty to eighty hours. Meridian routes may be used from North America, to South America, across the Antarctic continent to Australia and Africa, respectively, which would bring New York in touch with Australia in about 100 hours, and Africa in about 130 to 140 hours.

Cold is no impediment to air transportation but is a decided help because it freezes up the lakes and covers the ground with snow, both of which make landing for aircraft easy on the surface of the earth. Cold also renders the holding of moisture by the air difficult. This prevents the formation of fogs and clouds which lead to storms, heavy winds and uncomfortable conditions for flying. Aircraft should be designed to give great comfort to the passengers and security to any cargoes carried. Air transportation and air power, therefore are creating a totally new element in the relations of nations; this is principally a military one so far but soon its economic aspect will be even greater.

As physical means are employed by nations to impress their will on an adversary only when other means of adjusting a dispute have failed, air power will be called on as the first punitive element. The virtual extension of frontiers completely over a nation, which air power has brought about is entirely different from the old frontiers that consisted of coast lines, rivers, or mountain barriers, when armies and navies were the only factors concerned in military power. Now air forces may attack any town or hamlet no matter whether these be on the shores of the ocean, the crests of the mountains, or the inland regions of the countries subject to international dispute. This factor alone will cause nations in dispute to consider long and carefully the questions involved before resorting to an armed contest.

As conditions are at present no nations will willingly give up a reasonable organization for the defense of its territory, the maintenance of its institutions and the furtherance of its civilization either by civil, commercial or military means. On the other hand, due to the increased radius of operation of modern vessels and sea craft the conditions for the maintenance of bases on foreign shores are not as necessary as they were a century ago. Modern commercial vessels can go around the earth on one charge of fuel. Motor ships with internal combustion engines burning heavy oil are able to go with their full cargo for 25,000 miles. Each employs an incredibly small number of men to handle it. One of these ships of 15,000 tons requires a total crew of only about 25 men. The maintenance of militarily defended sea craft bases distant from the home country no longer has the strategic importance it used to. They often serve as means of irritation and fear to the nations that consider themselves menaced by them. It seems probable, therefore, that nations at the present time will be willing to enter into conversation regarding military establishments designed entirely to take the offensive across the seas and away from their own territories, especially where these have little or no defensive value. Many of the institutions and establishments still maintained by the nations for offensive military use across the seas have little or no actual value and have become obsolete and useless. Their maintenance requires excessive taxation and personal services in the building up of industries and classes which cannot be used for anything except war.

The three elements of national defense which nations will not give up until their protection is assured by some means which has not yet been tried successfully, are air forces, armies and submarines.

Surface navies, particularly battleships and other surface craft, are rapidly losing their importance as they can only be used for offensive operations across the sea and have little or no defensive value. They are the most expensive equipment in the military scheme.

What is keeping them up as much as anything else and largely preventing open and free discussion of their uses are the propaganda agencies maintained by navies for perpetuating the existing systems. Not only do they resist any change which will take away from the battleship its primary importance in sea dominion, but they tend to minimize and depreciate the ability of air power and submarines. Propaganda has a great effect on the public mind. As the people are the ultimate judge of what national defense should be, as expressed by their legislatures, it is necessary that the exact facts be submitted to them without restraint or evasion.

Heretofore armies and navies have had such well defined spheres of operation that they have been left largely to themselves to organize and carry out their own plans. Air power has changed this, and if air power is not given a sufficient voice in the national councils to be able to compete with the voice of the land and of the sea it cannot exert the influence commensurate with its power. This fact, among others, has led the various nations to give to air power an organization coequal with the army and navy and tend to group all means for national defense under one general head who can be responsible for all the armament of a state. This organization tends to distribute the duties of defense in accordance with the particular requirements of the case to armies, navies and air forces, to prevent any one arm from being expanded inordinately or from getting a dominant political position before the people and the legislatures by means of propaganda which may not exhibit a nation's exact needs.

There are three ways, therefore, of bringing on a discussion as to the limitation of armaments.

First, to show what is useless and can be eliminated in the national defensive armaments.

Second, to show what is of a purely defensive nature for the protection of the country itself and not necessary for prosecuting offensive war beyond the seas.

Third, what governmental scheme of national defense is the best suited to a balanced organization, so that air, land and water will each be represented, provided for and understood, and to bring about a full and exact publication of all arrangements and expenditures for the national defense.

The effect of air power on the armies is not as decided as it is on navies. Armies will still be used very much as they have been heretofore, primarily for insuring tranquillity in the countries themselves, for maintaining the constitution and the laws when other means fail, and last, for use against other armies. It is difficult to see in this country in particular, how armies will ever be used against other armies in the defense of the country if an adequate air force exists, because to get to this country hostile countries must come either through the air or across the waters. In either case an air force is an efficient and positive protection. An army can neither oppose an air force in the air nor a navy on the water. As everything begins and ends on the ground and as armies are the manifestation of the man power of a State in a military sense, they must hold the land bases from which air forces or sea forces act.

The people in most countries are so busy with their vocations that they have comparatively little time to look into the exact conditions of national defense. This is particularly so with nations that are far removed from an apparent possibility of future conflicts. They entrust their defense to professional bodies known as armies and navies, and pay comparatively little attention to what arrangements they make for national defense, providing they do not spend too much money or become too resistant to public opinion. This nonchalance always results in armies and navies if continued too long. It is manifested in the retarded development of modern equipment and the slow adoption of the latest instruments, and in a disregard of changed conditions of education, enlightenment and progress among people, which tends to settle international conflicts by other means than by war. The personnel of these permanent establishments often tend to become uniformed office holders instead of public servants entirely engaged in furthering the betterment of their nation.

As a preliminary step in developing the whole theory of the limitation of armaments, it is believed at this time that it is practical to do away entirely with the surface battleship, the airplane carrier, certain naval bases and dock yards, and many useless and expensive organizations of ground coast defenses. To bring this about, frank discussion, a truthful exposition of facts, and the widest publicity are needed. Our last conference for a limitation of armaments was greatly facilitated by the absolute proof that aircraft could destroy surface seacraft and battleships. Since the last conference in 1921 air power has made such strides that its effect is still more appreciated, and further limitations are now practicable.

We may look forward to the establishment of a permanent international committee charged with considering the views of any nation on the question of a limitation of armaments. Should any plans be thought practical of enforcement by this body they could be submitted, after discussion, to the countries interested in the limitation for their mutual benefit.

This need carry no obligation to follow the recommendations or the findings of an organization of this kind. Its object should be to expound to the people the exact value, uselessness or ineffectiveness of existing arrangements for national defense, their financial cost and their relative effect in international disputes. The people would then be in a position to judge about defensive provisions for themselves.

From time immemorial nations have striven to avoid war, to settle their disputes by treaties and covenants and to establish international tribunals to which appeal could be made when serious disagreements arose. These methods of adjusting disputes have gained considerable ground within the last generation, particularly with the coming of rapid means of communication and increased education among the people. The psychological attitude of the nations, due to increased education, is aiming towards not only settling disputes in an amicable way but also to eliminating causes of friction. Large and exaggerated armaments are and have been causes of apprehension and friction between nations, and as this phase of international competition seems a practical one to consider in some of its phases, the nations should be willing to discuss certain features of a limitation of armaments program.

To make a limitation of armaments effective, there must be a mutual and honest agreement among the powers. There must be nothing hidden or concealed about the methods of carrying out the treaties for disarmament and an adequate inspection of the equipment affected by the agreements must be made mutually possible by the powers involved so that there will be no occasion for distrust of the action being taken. Such limitation of armaments cannot compromise a nation's independent existence. It must be of practical and economic benefit, and it must be thoroughly understood by all people; otherwise, the provisions will fail, they will lead to distrust and may even lead to a greater competition in armaments than exists at the present time.

A—The 1100 pounder that sent the " New Jersey " to the bottom
B—The " New Jersey " turned over. (1923)

The Effect of an 1100-lb. Bomb on a Battleship

(1)—1100-lb. bomb hits.

(2)—The appearance of the ship 6 seconds after the impact.

(3)—The appearance of the vessel 1 minute afterward. All masts are shot down, the decks cleared, and the ship is destroyed. This single shot, with the other bombs hitting in the vicinity, caused the battleship ***Virginia*** to sink within a few minutes. All battleships, no matter how constructed, can be destroyed from the air in a short time.

Spraying Grove of Catalpa Trees to eradicate insects and disease

This is a successful way of combating these conditions

U. S. Army Airplanes Refueling

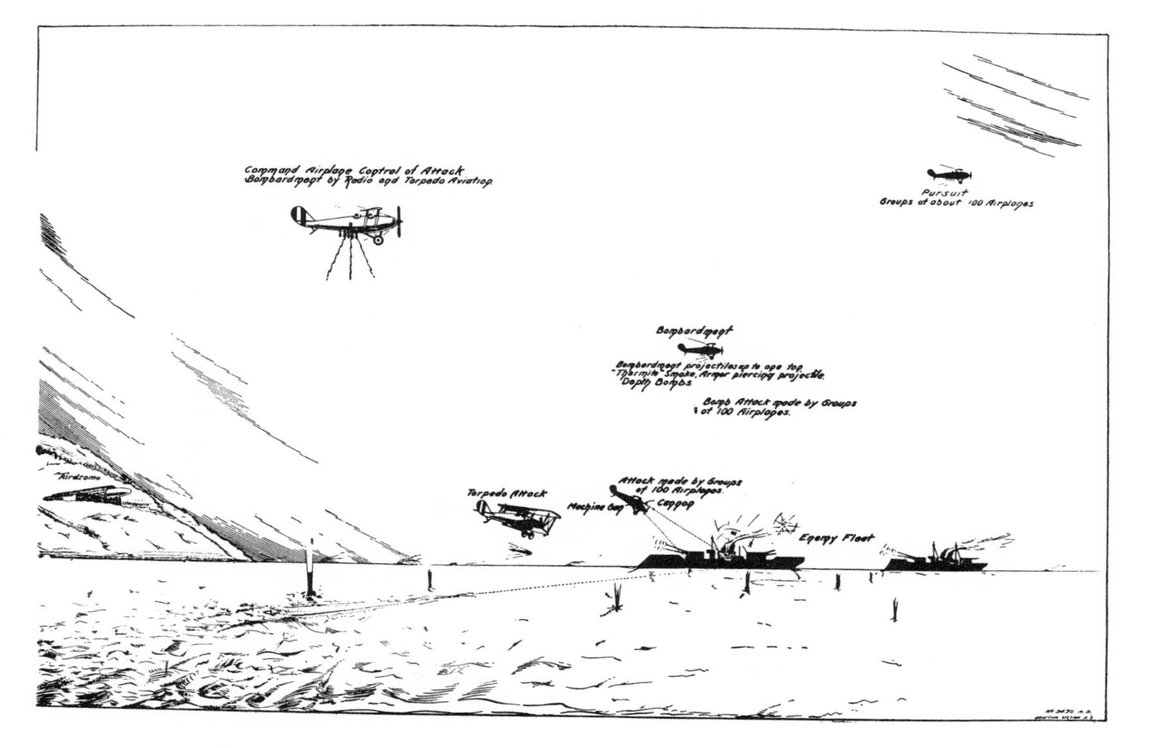


Diagram—Principles governing functions of Air Service operations over water. Day attack on Hostile Fleet. No. 6

The attack is made from above, gliding and on the same level with torpedoes

Forest Fire Plane on the Pacific slope

This saves millions of dollars annually

Lieutenant Macready in Le Pere supercharged plane. He rose to an altitude of over 30,000 feet

General Mitchell and Colonel Milling visiting the French Army in the advance lines on the Champagne front, September, 1917

Note German shells bursting in the background

Parachute descending

Many officers have been saved by this device when their airplanes have broken

Three-inch trailer mount anti-aircraft gun in action at night. Searchlight in the background

Fire from ground guns has very little effect on aircraft

Smoke (Gas) Curtain being laid over New York City

Whole areas can be inundated with gas in this manner

Smoke screen over Langley Field, Va.

This is used to cover troops or battleships while the bombing planes attack

General Mitchell in conference with Congressman Lampert, Chairman of the Aircraft Investigation Committee, U. S. House of Representatives

President Coolidge being shown General Mitchell's plane by Airplane-Inspector Harry Short

For Waking Him Up -:- -:- By T. E. Powers
GEN. MITCHELL
BOSS, YOUR HOUSE IS ON FIRE!
YOU'RE FIRED! FOR WAKING ME UP
THE FLYING AGE
THE ARMY BERTH
WAR DEPT. BUREAUCRAT
NAVY DEPT. BUREAUCRAT
TESTS HAVE PROVEN THAT THE AIR PLANE IS THE BEST MEANS OF NATIONAL DEFENSE
AIR SERVICE
WE DON'T CARE MUCH FOR THESE HIGH FLIERS
CRITICISM
GEN. MITCHELL
U.S. AVIATION SERVICE
ARMY BUREAUCRAT
ARMY
NAVY
DUCK WITH THE UNTRIMMED WINGS
AIR SERVICE
ADMIRAL SIMS
I'LL TELL THE WORLD THAT AN AIRSHIP CAN SINK A BATTLESHIP!
SEC WEEKS

"Sailor Beware!"
BRIG GEN MITCHELL
US NAVY
ASSERTION OF AIR-CRAFT SUPREMACY
SOS
U.S. AVIATION
ARMY MULE
NAVY GOAT
WAKENING OF RIP
IT'S STRANGE THO BECAUSE I INVENTED FLYING
INADEQUATE
AIR SERVICE
WAR DEP'T
ARMY
NAVY
AIR SERVICE
ARMY
GEN WILLIAMS
MITCHELL
ADMITTED AIR SERVICE DEFECTS
MITCHELL
ARMY
NAVY

## VII

## A GLANCE AT MODERN AERONAUTICS

THOSE of us who have chosen the air for our profession naturally are enthusiastic about it. We know that the essence of civilization, of communication, of national defense, and of all development, is transportation. Transportation with us is not a question of land, or a question of water, or a question of mountains, or a question of deserts, it is the air, and the air permeates everything.

We are not limited to the air near the surface, but have climbed to altitudes of forty-four thousand feet and can go still higher. Our routes through the air are limitless, the only thing that curtails air transportation being the fuel capacity and the reliability of our engines. The fuel capacity is being increased every day and the percentage of engine failures is rapidly decreasing. I can say now, definitely, that we can encircle the globe in a very short time on a single charge of gasoline.

Aeronautics, as you know, is a very recent development. The conquest of the air was made possible by the development of the internal combustion engine or the gasoline engine, as we know it. The present type gasoline motor with its cylinders, its connecting rods, its crankshafts, camshafts, and all that multiplicity of gears, is still a complicated mechanism. We feel confident that a better means of motive power will be developed which will be much lighter and more reliable. This will again increase our ability to go through the air.

The rapid development in aeronautics after the first successful flight by the Wrights is directly attributable to the war. This was due in a large measure to the fact that aeronautics is the only instrumentality for delivering something at a terminal station in the air—whether it be the eye of an observer, a machine gun, or a bomb. Therefore, it took its place at once as a military instrument of prime importance.

Aeronautics was first employed to deliver an observer who could report back enemy activities. Next, machine gun fire was delivered from the air, and last instruments were devised for demolishing things on the ground. It was definitely established during the war as an absolute principle that if you did not have sufficient control of the air to be able to operate, your ground force could not carry on against the enemy who had supremacy of the air. In fact, before the close of the war, if either side had been deprived of its aviation service, the opposing army certainly would have won within a few weeks. There is not a question of a doubt about it.

After the advent of aviation, troops could not move in the daytime without their presence becoming known to the enemy; an attack could not be organized and carried through without the absolute co-operation of the air force.

The war in Europe was essentially a conflict to be decided on land. There were many things connected with it that depended on water transport, still the ultimate decision was being sought on land because Great Britain dominated the sea to such an extent that her supremacy was not seriously questioned except by the submarine. Aviation was developed, primarily, for use on land and for short distances from an operating base. The average distance from an air-drome was about sixty miles. The equipment, the training, and the methods of employment were devised for this particular situation.

Since the war, development has been continued and aviation has been successfully applied over the water. Some very interesting things have been discovered over this element. There is the practical employment of the theory of the "water hammer," that is, the utilization of the power of water itself as a striking medium. On account of the compressibility of the air, we soon reach the limit of the amount of explosive that can efficiently be employed for destruction in demolition bombs. If, for instance, we were to explode five hundred pounds of TNT in a city street, we would create a certain amount of havoc. Increasing the charge above that amount would not produce a very great difference in the amount of destruction that would be wrought. On the other hand, if we were to explode TNT in water, a medium that is practically incompressible, we obtain altogether different results. In the air, we secure only a slight increase in effectiveness by doubling our charge of explosive, but, if we double our charge of explosive under water, we get not two but perhaps four times the effect. It is terrific. In other words, when water is impelled by these great charges it has practically the striking force of steel. If it hits the sides of a ship it smashes all the way through. The development of aeronautics for employment over the sea has progressed as far as its utilization over land and it will probably have more effect in the control of water areas than it has ever had over the land.

Since the war we have sought to develop our equipment for the purposes of transportation and the economic development of the country. Commander Christie and I went, in February, 1923, to Camp Borden, Canada, having flown there from Detroit. The cold was intense. Our airplanes were equipped with wheels when skis should have been used on account of the deep snow. As a result, the machine was thrown forward on its nose but no serious damage was done. From Camp Borden we proceeded to the railroad station by horse and sleigh, a distance of about eight miles. It took us as long to cover the eight miles by sleigh as to traverse the two hundred miles from Detroit to Camp Borden by air.

One horse, along that trail, could pull about one thousand pounds and could move it at, we may say, a maximum of twenty miles a day. Now, in the air, we could make one hundred horsepower lift one thousand pounds and go four hundred miles a day easily. It is possible to go one thousand miles a day. It is not uncommon for me to have my breakfast in Washington, my lunch in Dayton and my dinner the same night in Chicago, Detroit, or Milwaukee. There is nothing remarkable about it. I do it as a matter of fact and do not consider it at all extraordinary. The question of cost is another consideration. It costs from one to two dollars a day to keep a horse, while with the antiquated war machines we are flying it costs about sixty cents a passenger a mile to fly. It cost approximately a dollar and twenty cents a mile for Commander Christie and me to fly about two hundred miles from Detroit to Camp Borden. We covered the distance in one hundred minutes. Our plane was a war machine with a great excess of horsepower. It generated approximately four hundred horsepower when thirty or forty horsepower might have carried us just as well. With one hundred horsepower in the air we can lift and carry more than a horse can pull along the ground, and can move the load about thirty times as far during the day.

If speed is essential in the delivery of your goods you may obtain it through the air. In connection with air traffic you have probably heard a great deal about accidents. We shall always have accidents in military aviation, because in military aviation we require the greatest speed and the greatest possible carrying ability. We must fly in all kinds of weather and must land in very unfavorable places for the proper prosecution of military operations, and it is a principle with us that we fly day or night, regardless of weather conditions.

We accept this in military aviation as a general principle, but of course, there are times when we are stopped. On the other hand, with commercial aviation we can organize regular air highways, provide meteorological facilities which will give definite indications of the weather, and install emergency landing fields all along the routes over which we intend to operate, so that practically we should have little loss. In fact, some of the operating companies in Europe say they have had very slight losses in the last four years. Their record of operations is much better than that of the railroads for a corresponding number of passengers, quantity of freight hauled and schedules maintained. Statistics will bear out these conclusions.

There are many ways of applying aviation to civil pursuits. It is invaluable as a mapping agency. In the United States, for instance, we have little more than forty per cent of the country mapped. We have been making surveys at great expense for a great number of years. With two hundred airplanes properly equipped we can map the whole country, hitch up every bit with the established control points, and do the entire job more accurately than can be done by any other means. It could be mapped by this method for from one-tenth to one-fifth of the cost necessary by other methods, and the whole project could be completed in about two years.

In 1919 we assigned one squadron of fifteen ships to duty with the Forestry Service for the purpose of conducting fire patrols, and in that summer alone—1919—the Agricultural Department reported to the

War Department that we had saved more money to them than the total sum expended that year for all Government aviation. This is the more remarkable when you realize that it was the first time we had utilized an airplane forest patrol.

We do a certain amount of soil survey work for the Department of Agriculture. From the color of the soil in certain localities and from the character of vegetation at certain times of the year it can be determined from aerial photographs what fertilizers are necessary, and what crops can be raised to best advantage. We can ascertain the amount of improvement to farm lands, conditions of the crops, the drainage, water supply, areas of blight, and many other conditions affecting agriculture. You would be surprised if you should study the photographs to see how accurately these things stand out. Such surveys act as an incentive for one farmer to compete with another, as they show graphically what results are obtained by different methods. We have done some very interesting work along these lines pertaining to dairy farming in certain parts of this country.

Some of our fruit trees have been attacked by various parasites and we have successfully sprayed the trees from the air and eliminated the insect and parasitic pests. Locusts and boll weevils can be destroyed. This is all a matter of record. We can tell you what it costs, how long it takes and the best method of procedure.

At the present time, we are studying the problem of the transportation of goods, with particular reference to the transportation and delivery of mail. It now takes about five days to deliver mail from New York to San Francisco if transported by rail. The distance is about twenty-four hundred miles. About four years ago we opened up an air route from New York to San Francisco. This was done by running a race against time from New York to San Francisco and return. The total distance was approximately five thousand miles. We made the trip from coast to coast in twenty-four hours flying time and the return trip was completed in about one day's flying time. After that a certain amount of development work on the airdromes was done and a control system installed. The Post Office Department began to carry mail by air, at first only in the daytime, transferring it to the mail trains at night. Later, a part of the highway was lighted for night flying, that is, from Chicago, Illinois, to Cheyenne, Wyoming. This part of the route is now traversed at night. They operate on a thirty-hour schedule from New York to San Francisco. The whole distance is now being lighted and within a short time a twenty-hour or less schedule can be put into effect. I believe that 95 per cent of the trips will be made on schedule time, and, if properly organized, that there will be very little loss. The mail will leave New York about midnight, will be delivered in Chicago about five in the morning, will fly by day to Cheyenne, Wyoming, and reach San Francisco in the evening for early morning distribution. A passenger and freight service will follow.

In all our operations of aircraft we have done comparatively little work in flying where the temperature is really low. However, we do not believe that there will be any difficulty in solving cold weather flying. Great progress has been made in Alaska and Canada. We have operated successfully at 60 degrees below zero. It is just a matter of development. The difficulties that bother us most at ordinary temperatures are storms and fogs near the ground. Flight in aircraft has given us considerable knowledge about atmospheric conditions that was not available previously. For instance, in the Chesapeake Bay area in which we have done considerable operating, we have learned a great deal about the conditions that cause local storms. When the sun comes up in the morning the weather is generally very clear. As the atmosphere heats up, the sun pulls moisture up into the air, because the warm air can hold more moisture than cold air. This moisture gradually begins to rise and this movement continues throughout the day. The Appalachian Mountain System, lying but a short distance to the west, is covered with trees and vegetation. The vegetation absorbs the heat while the temperature of the ground remains fairly low. Cooler currents of air start moving down the mountains towards the Bay. As soon as the sun starts to lose its heat in the afternoon the atmosphere begins to cool. The currents of cold air coming down from the mountains strike the warmer moisture-laden areas, which causes the moisture to congeal and we get cirro-stratus and cirro-cumulus clouds. As the atmosphere cools down the air currents begin to change with a corresponding change in wind direction, that is why it is a common saying that "the storm works against the wind." As the clouds move they become charged with static electricity. That causes lightning. As the air cools down a thunder storm begins to form and travels in a direction opposite to that in which the wind is blowing. Such storms are usually twenty miles in diameter and travel from twenty to twenty-five miles per hour along a fairly definite route. We have observed these storms until we can determine fairly accurately what their path will be. Although the storm as a whole may move at twenty miles an hour the wind velocity within it is sometimes one hundred and fifty miles. As the storm moves along, the air near the ground is retarded by coming in contact with trees, houses, and other obstacles, in just the same way that the current along the bottom of a river is retarded. In consequence the air above travels faster and the result is that the whole thing moves along the ground like a rolling cylinder. This downward rolling movement of the air has a tendency to precipitate an airplane onto the ground in the same manner that a current may precipitate a ship against the rocks. People used to believe that they could fly over such storms, but sometimes we find that they go up as high as the atmosphere will hold moisture. Scientists estimate this to be about fifty-five thousand feet or about ten miles. With the equipment available at present the best thing we can do is to keep out of these storms. They are plainly visible and it is possible without difficulty to go around them, between them or to avoid them entirely.

Aircraft need have no great difficulty with such obstacles. All that is necessary is a proper meteorological service equipped with an efficient radio system and good navigation instruments on the planes. A good radio system is absolutely imperative in order to carry on. Little can be accomplished without it. Communication along the ground is too slow. In order to operate transportation through the air, communication must be through the same medium.

In Europe it is necessary for the Powers to organize their air forces so that they can be made immediately available in case of war. In Europe, also, where the frontiers of one country directly abut on the other without any natural boundary, it is found to be absolutely necessary for them to organize their air traffic. Air power depends, among other things, upon the proper organization of air routes, and as air routes are absolutely essential, European countries have found it more economical to subsidize commercial companies even up to fifty per cent of their entire cost to carry on this work than to have to pay the whole thing themselves. These air routes will be utilized by military aircraft in time of war and the commercial craft will be converted immediately into offensive weapons. Commercial aeronautics always provides crews of trained pilots, while the airdromes are capable of being used as concentration points during war. Commercial competition will develop the whole science of aeronautics more rapidly than is possible by Government development work alone. For these reasons European countries have encouraged the establishment of civil and commercial aeronautics to the greatest extent practicable.

To illustrate the principle of subsidy: If a company desired to start a line from one point to another and would supply suitable equipment which would come within certain requirements imposed by the Government to make it useful in case of emergency, the Government would pay one-half of the initial expense of the operating equipment. If the ships are actually utilized in commercial traffic the Government provides a certain subvention to the companies based on the number of pilots employed, the number of mechanics, number of passengers carried, amount of freight and the speed. It all amounts to an assistance of about fifty per cent. In some countries they guarantee a five per cent net profit on the capital invested. This is the maximum guarantee. If over five per cent profit is realized, the Government deducts the amount from the subvention.

The requirements imposed on aircraft are such that they could be immediately adapted to war uses upon the outbreak of hostilities. Should war be declared in Europe it would take at least two weeks to concentrate land forces on any front. It is necessary to concentrate the personnel at mobilization points. They must be conveyed by rail, motor, or marching to the theatre of operations and then they must be deployed for action. Airplanes, on the other hand, can take the offensive at once. During the next war airplanes will not be limited to a radius of action of sixty miles; we may expect to have them operate several hundred miles from a base. The European air forces are co-ordinating with civil air transportation and with every means for putting civil air power on the offensive instantly. Every ship goes right into the offensive at once. That is the basis of their whole organization.

Air ways may be established in the arctic regions as well as in the tropics. In the north, during the winter time, nature provides an airdrome of snow and ice, everything is smooth. The interstices in the ground are filled up with snow and the rough places are made level. The lakes are frozen over with ice which provides a substance hard and smooth for a landing field. Airplanes can land practically anywhere except in the timberland. One can find water courses, lakes, and very practical landing fields almost everywhere. Air transportation is the best kind for getting from place to place in the far north.

I remember going into the north, into our Alaska, many years ago. We were building a telegraph line in the Upper Yukon. It was generally believed that the work could be carried on only in the summer time, because it was too cold to work in the winter time. As a result, little progress was being made as the pack animals sank into the moss up to their knees and could carry only about two hundred pounds on their backs. In the winter, when the ground was covered with snow and ice, the load could be raised to two thousand pounds and could be pulled much more easily. By changing our system and working throughout the winter we succeeded in getting the line finished. I do not believe that it could have been done otherwise.

I believe that the development of transportation in the north will occur in the winter time. Heretofore the use of gasoline engines has been limited to warm temperatures, but I see no reason why, if we work hard on the problem, we cannot get them to work satisfactorily in cold weather. It is only fifty-two miles across the Bering Strait to Asia and there are the two Diomedes Islands in the middle, only six miles apart. The widest stretch of open water between North America and Asia is only twenty-one miles, and in the winter time that is pretty well filled up with ice.

On the other hand, in going to Europe by way of Greenland and cutting across to Iceland there is no place where you have clear water for a greater distance than about three hundred and fifty miles, which represents about three hours of flying time. Aerial communication with Europe is certainly possible along the northern route if properly organized.

In the increase of geographic knowledge and in the development of exploration, aeronautics can play a very prominent part. You can get more out of aeronautics than out of any other means of communication because you can go straight through the air to your destination. All that is necessary is to have the proper organization. It is impossible to start out with one ship or two ships and expect to do the thing at once, but if the route is intelligently organized ahead of time, it will be possible to maintain communication. Aircraft can supply themselves by air, if necessary. In case it is necessary to make a forced landing, it is possible to communicate with the base by radio. Development work has been done along that line. As a matter of fact, we have carried out exercises during the winter in which we have gone out with our squadrons, established airdromes in the snow and then supplied ourselves by air. We used a transportation plane carrying thirty-three hundred pounds, that is, one and one half tons of cargo. It is possible for one ship to make two trips from a base at a distance of two hundred and fifty miles from the operating units on the same day, or double it with planes of greater radius.

I want to say something about the lighter-than-air machine, that is, the dirigible. There is really only one country that has gone into this question thoroughly. Lighter-than-air machines have been used for a great number of years. The French took a balloon with the Revolutionary Army into Belgium in 1792. Napoleon used balloons in Egypt. Balloons were again used in military operations in our Civil War. It was only after the coming of the gas engine that a beginning was made in propelling them through the air. The Germans used a dirigible airship in 1900. There were many ups and downs encountered with it, but before the World War the Germans succeeded in carrying 200,000 passengers in airships without a single accident. Our figures show that in the larger airships, with a full load, passengers can be carried for distances of over five hundred miles at about three and one-half cents per mile. We know also that other airships can be towed through the air. This, of course, would increase the loads accordingly and reduce the motive power required.

In discussing the practicability of using airships the question is always asked about the cost of terminals. The Grand Central Railway Station in New York with its terminals, cost about $200,000,000. The Lake Shore Station and its terminals in Chicago cost about $60,000,000. Our little station in Washington with its terminals, cost $30,000,000. The cost of an airship station as compared with those is very small. In the operation of airships we do not have the upkeep of tracks and other expenses incident to the operation and equipment of railroads. An excellent lighter-than-air station can be constructed, capable of housing five airships, for about five million dollars. Five airships running continuously from New York to Chicago could carry as many people as are carried on all the fast trains at the present time.

The British have done some very good work in the development of aerial aircraft carriers, that is, in having heavier-than-air ships land on, and take off from, airships. They have never actually effected a landing but they have done practically all the necessary experimental work up to that point. They would have done this had appropriations not been curtailed. There is nothing complicated about it whatever. We took it up where they left off and have landed an airplane on an airship without difficulty.

Satisfactory progress is being made in the development of helium gas. By using non-inflammable helium gas in airships, a means of transportation is developed which is sure, safe, and reliable.

We have, then, two means of navigating the air that are fairly well developed. One is the airplane, which depends upon power within itself to carry it along, that is, it has to come in contact with just so much air in a given time to gain sustentation. The more power it has and the smaller we can make it the more speed can be obtained. If we desire a greater carrying structure we build a plane with more surface. This, however, requires a longer run on the ground to accumulate the necessary speed to generate lift, and conversely, a relatively larger space in which to land. Then we have the lighter-than-air machine or airship. This requires a medium lighter than air for its lift. Neither of these classes have reached their ultimate development but real progress is being made daily in perfecting them.

We are developing a third means of aerial locomotion which will depend entirely upon mechanical means of ascension and which will lift and land itself vertically. This is known as the heliocopter. People have tried for a great number of years to solve the problem and a great many theories have been experimented with, all of which have, invariably, been unsuccessful. Within the past two years a practical heliocopter has been developed which has risen vertically and travelled from one point to another. When the heliocopter is perfected it can be used as a feeder for other air lines. Airplanes require a fairly large landing field while the heliocopter, due to its ability to come down vertically, may be landed in a very small area. We believe that, from both a military and commercial standpoint, the heliocopter eventually will be of great value.

Another development which has been much talked about recently is the motorless airplane or glider. Gliders depend on rising air currents for their sustentation. Rising air currents are produced in a great many ways. You may get an upward air current from the wind impinging on the surface of the ground. You have all seen the various gliding birds moving around in the atmosphere taking advantage of the air currents. You have all observed the little air currents blowing leaves and picking up small articles and resolving themselves into small whirlwinds. We are learning a great deal about these things, which is increasing our information on atmospheric conditions and which will have a tendency to produce better aerofoils or wing sections for our airplanes. We are trying to encourage glider development in the United States. Gliding is an excellent sport, is good training for pilots, is inexpensive, and will probably do a great deal for the development of aeronautics.

The sum and substance of all this is that military aviation is a thing that a nation must have as a defensive arm. Without a well-organized air force the nation is helpless to-day against one that has an efficient air force. Commercial aviation has been developed to the point where it now costs, mile for mile, about three times as much to travel by air as by rail or boat. The cost has been reduced three times since the war and is being further reduced all the time. In figuring the cost, as stated above, the time element has been totally disregarded. There are certain parts of our country where it would pay to transport material by air. You might figure out, for instance, what it costs to transport a certain article from the mouth of the Mackenzie River on the Arctic Ocean to the point where it is placed on the railroad train; what the depreciation is on it; and the amount of interest on the money that is tied up in it for a definite period of time due to slow transportation. I refer to precious furs, for instance. I believe that flying equipment can be devised to bring furs from the North more economically than they are being carried at the present time. There would be a greater certainty of delivery through the air than by the existing means of transportation, dog teams in winter and canoes in summer. However, in preparing for commercial aviation it is always better to make a survey for it ahead of time, so that you will know exactly what you are going to do. Instead of building airplanes first and then trying to find things to carry in them, the procedure should be reversed. It would be absolutely foolish to build a railroad from one point to another in a desert when you have nothing to carry on it. The proposition seems simple on its face, but commercial aviation has done that very thing in the past, which is one of the many reasons why it has not been universally successful.

I have tried for a long time to get our Government to make a survey within the country which will show the people what they can transport advantageously and economically by air. For instance, some mail order houses have to utilize three or four means of transportation, motor, railway, animals and ships to reach their purchasers. I know that, in some cases, full loads delivered by airplane would be more economical than the present transportation methods.

When aerial transportation has been carefully studied by people who understand the operation of aircraft, it can be placed on a successful economical basis. It is bound to develop. In fact, it is developing probably more rapidly than any other means of transportation at the present time, so that within a few years we may expect air lines from the United States to Europe and Asia via the Arctic and to Australia and Africa via South America and the Antarctic Continent.

## VIII

## THE MAKING OF AN AIR FORCE PERSONNEL

MEN and machines have to be harnessed up and driven as a team to make up air power. The selection and training of the persons who are to fly the machines and those that are required to keep them up is the most important consideration. The next is to obtain and distribute the actual airplanes and the equipment that are necessary for use in the air.

Without knowledge on the part of the personnel of their work, neither proper air units nor suitable material can be devised or created for the flyers. If persons are put in authority that are not trained air officers, with long service as pilots and observers, they cannot know the kinds of airplanes which should be given their men and the material which should accompany them to keep them up.

All countries have attempted at first to put men in the control of aviation who knew little about it just because they had high military rank. These officers always attempted to conceal their ignorance of the subject from others, and have surrounded themselves with advisers that knew little more about aviation than themselves so as to maintain greater control over their subordinates. The result of this procedure always comes quickly and is manifested in worthless and dangerous machines for the pilots, an inadequate system of training, no real air system for reserve officers, and no appreciation of what the conditions of a future war will be. Everything depends primarily on the creation and development of a specialized air personnel, capable of actually handling their duties in an efficient manner, making a class of real air men.

An air force's duty is in the air and not on the ground. People who are unused to or unfamiliar with air work are incapable of visioning what air power should be, of training the men necessary for work in the air, or of devising the equipment that they should have. The greatest deterrent to development which air forces combat in every country is the fact that they have had to be tied up to armies and navies where senior officers, unused to air work, were placed in the superior positions at the beginning of the organization of the air forces. In practically all cases these affected to treat flying men as aerial chauffeurs, where as a matter of fact, they are the most highly organized individual fighting men that the world has ever seen.

Airplanes are not merely a means of transportation, they are fighting units. Air forces fight in line against other air forces. They use their own tactics, and have a highly specialized method of maneuvering in three dimensions. The air man's psychology of war depends on the action of the individual, he has no man at his elbow to support him; no officers in front to lead him, and no file closers behind him to shoot him if he runs away as is the case in a ground army. The whole system is entirely different from that of troops on the ground where mob psychology has to be used in directing the men in combat. To cover up their ignorance in these matters, these older ground officers have always hedged back to the fact that administration was the main thing in the conduct of air forces. Administration is merely the orderly conduct of correspondents in affairs. It has nothing to do with the actual handling or leading of fighting forces. It is merely a necessary nuisance. The best administrators usually are the old sergeants or clerks that have been long in the service. An excellent administrator could be obtained and hired for certain fixed wages in civil life. An airman cannot be. He must be of suitable personal characteristics, self reliant, bound to overcome any and all obstacles in front of him, and well versed in his profession from the ground up.

Another thing that one frequently hears is that the air game is a young man's game. This is not the case when one considers that it is a life's work. Even in a ground army, a general would no longer be capable of carrying a musket and pack in the ranks nor making the long marches on foot that he used to when twenty years of age.

An officer has to come up through each grade to his position, and in addition he has to learn about all the branches of the service in the army to be a general. It is the same way in the air force. One starts as a pilot, learns how to fly the machine, then learns how to handle a greater number of machines as rank and experience increases, and last, to handle whole forces of all branches of aviation, including their supply and upkeep.

It takes much longer to train upper officers in an air force than it does in a ground army or navy. To begin with, every boy has the background of an army and navy even in his primary school. He is marched round, and has some instructions in military movements. He hears his parents talk about wars, and studies about them in the school histories. He goes to the sea in ships as well as in boats, and also reads about sea combats. He is more or less familiar with both. The modern boy is the first of his kind to know about the air. He is just beginning now to get a background of the air, but, unfortunately, air battles and air history of the European war have not yet been written. The ground troops were so busy with their own work, as they always will be, that they knew practically nothing of the great air operations that took place miles behind the enemy front or sometimes miles behind our own front. Air forces act in such an incredibly large space. They are capable of going one or two thousand miles a day so that the whole country becomes a frontier—a horizontal frontier,—with the ground being one element and a blanket of air all over it the other element. It is not a vertical frontier like the coast, a river line or mountain barrier. Speed of locomotion is tremendous in the air. It is ten times as fast as the average steamship, four times as fast as the average railway train, and nearly a hundred times as fast as the foot soldier. Furthermore, an air force fights in three dimensions, on the same level, from up above and from underneath. It can go wherever there is air and this air is about seven miles deep. In other words, fighting air craft can mount to 35,000 feet. The control of all the different branches of the air service in this medium of air is a most complicated and difficult undertaking. The technical part of it alone is more complicated than that of any other service. More than seventy-five different professions are necessary to be known by the personnel handling the aircraft to keep them in repair and in serviceable condition.

In the actual fighting of the aircraft, moral qualities are required that were never before demanded of men. In the first place, they are all alone. No man stands at their shoulder to support them. They know that if a flaming bullet comes through their gasoline tank it immediately becomes a burning torch and they are gone. They know that if a wing is torn off there is the same result. They know that a dozen fatal things may happen anytime, and that if they fall two hundred or twenty thousand feet, existence is at an end. A man on the ground may be wounded and yet may be saved, as he falls a foot or two to the ground. In spite of all these things, the airman has to push his attack without other thought than the destruction of the enemy. Human beings that are endowed with these characteristics are not to be found without careful selection and elimination.

There are three great branches of an air force.

First, pursuit aviation, in which there are single seaters with only one person in each airplane. These are designed to pursue hostile aircraft, catch them, force them to combat and destroy them. The greatest qualities of individual daring, resourcefulness, coolness and physical ability are required in this branch.

It is upon pursuit aviation that control of the air depends. Pursuit aviation is the main fighting line of an air force. Its main object is to meet the pursuit aviation of the enemy and to vanquish it, thereby establishing control of the air. It is what infantry is to a ground army. In fact an air force must be able to defeat the hostile pursuit aviation or everything else will fail.

Another great branch of an air force is bombardment aviation. This is designed to destroy objects on the ground or water by hitting them with projectiles, or covering them with chemicals. They use the most powerful weapons the world has ever known. Their bombs now weigh up to and over 4,000 pounds. Much greater ones can be devised and carried, if necessary. They use water torpedoes of a similar nature to those carried by torpedo boats in the navies, and of course can launch them with much greater effect than is the case with seacraft. They use gliding bombs which can be directed to their objectives from a distance; that is, a bombarding airplane can launch a bomb which has wings and glides. Its course is controlled by a gyroscope which acts on its rudder so that it will go in the desired direction. It also may be controlled by radio and dropped where it is desired.

Bombardment aviation may use the aerial torpedo. This is really a small airplane with its engine and all control surfaces, but without a human pilot. It is directed by gyroscopic controls and can go as far as its fuel will allow. Over one hundred miles may be covered and it can be guided by the plane that launches it.

It is possible for an airplane to fly along and control by radio several other airplanes which have no human beings in them and which may be made to drop their bombs on a city. The way we try this out in the time of peace is to put these automatic controls in an airplane with a pilot in it to take hold of it in case anything fails, so as not to smash up the airplane or cause destruction to anything on the ground by hitting it. These airplanes are equipped with cameras that are snapped by the device that would drop the bomb. The photograph indicates where the bomb would hit.

It is no longer necessary for aircraft to fly over a site or district to be able to hit it by projectiles. They may stand off several miles and do sufficient damage to cause a city to be evacuated. Chemical weapons of all sorts may be used. These are what the world stands most in fear of at the present time.

The progress made since the war in bombardment aviation has been phenomenal. The sights which are used now are very accurate. Airplanes can launch their projectiles with great accuracy from practically any height. Bombarding airplanes also act in great numbers.

During the war we had to fly a bombarding formation directly over the objectives at low altitude in order to make hits. This gave the opposing pursuit aviation favorable chances for attacking them and our losses consequently were very large. The anti-aircraft artillery even then had very little effect. It never stopped an actual attack. Bombardment aviation always had to be protected by pursuit aviation. When the pursuit aviation engaged in combat, it often had to maneuver so as to leave the large and heavy bombardment planes alone in order to maintain itself against the enemy. Pursuit aviation climbs and dives, turns and zooms and in a minute or two may be several miles away from where it was. Bombardment planes, on the other hand, are unable to maneuver as rapidly as pursuit planes. They stick as closely together as possible, practically wing to wing, and depend for their protection on the volume of fire from either machine guns or cannon. The result has been that the enemy often used two bodies of pursuit aviation. They would draw our pursuit away from the bombardment with one body and directly attack the bombardment aviation with another body. Many instances of this kind occurred in France.

I might mention one happening in one of our battles at Conflans during the St. Mihiel operations. At that time we had the greatest air force ever employed under one command. We were bombarding the German centers of concentration behind their lines with all the force that we had, so as to interrupt their movements and supply, and force their pursuit aviation to the defensive to keep them away from our ground troops. On September 14, 1918, one of our bombardment squadrons belonging to a French group failed to meet the pursuit aviation that was detailed to protect it on account of poor visibility in cloudy weather. It therefore proceeded in the direction ordered to bombard the objective. There were eighteen airplanes in the squadron, fifteen being 2-seaters and three of them being 3-seaters. The 3-seaters were equipped with six guns each, and, as far as volume of gunfire was concerned, were the most powerful airplanes on the European front. They were unable to maneuver as rapidly as the single seaters, however, and therefore did not fulfil the ideas of their originators who thought that through volume of fire alone they could defend themselves against small, highly maneuverable single seaters. The 3-seaters were supposed to be for the protection of the 2-seaters, that is, these powerfully gunned airplanes were expected to fight off the enemy pursuit while the bombers that they were protecting could concentrate their whole attention on dropping their bombs on the targets.

The squadron flew in a V formation like a flock of ducks. One of the great 3-seaters was on each flank and one in the opening behind. When this squadron crossed the line on the way to its objective, it was passed by a patrol of twelve German pursuit airplanes flying one behind the other about five hundred meters above it. The German patrol deployed in line formation behind the bombardment squadron. Four of the enemy airplanes attacked the 3-seater which was behind and sent it down in flames. The other eight kept up a long range fire at the squadron so as to derange its aim while dropping its bombs on the city of Conflans. At the same time, anti-aircraft artillery opened fire at the vanguard of the squadron while the German pursuit ships attacked the rear. While anti-aircraft guns failed to hit any of the airplanes their bursting shells allowed the German pursuit organizations, which were now concentrating for an attack on the squadron, to see where they were. During this time, the Commander of the bombardment squadron noticed German airplanes rising from the airdrome close to Conflans, that is, at Mars la Tour. The bombs were all dropped on the objective and the return flight was started to our own lines. Just as the turn was made, a fresh enemy pursuit squadron joined the former, immediately deployed and attacked the rearmost plane and shot the observer through the leg. He continued the battle, however, and hit one enemy plane which fell in flames. The formation was now well on its way back when a third enemy squadron attacked ours in front and to the left. The bombing squadron was now being attacked in three dimensions, from underneath, up above and on the same level.

To one who has never seen a fight of this kind, it is impossible to convey an idea with mere words. The great lumbering bombing machines huddled together as a flight of geese might when attacked by falcons. The pursuit airplanes diving at them from all directions, firing their machine guns, then zooming up in the air or turning over on their backs at a speed of about two hundred miles an hour, taking an erratic course to avoid the fire of the big ships and then resuming their position for attack again. Frequently an airplane is hit, bursting into flames, losing a wing, or having its controls shot to pieces, or its pilot is killed instantly, when it spins away on its course to the ground, leaving a long trail of flame and black smoke behind it.

By this time the big 3-seater protection plane on the left had been shot in one of its engines and started slipping down. Immediately when it left the formation it was jumped on by three German machines. In a moment it was shot to pieces and disappeared in flames. Fighting now had become terrific. More German machines were constantly joining their comrades. The signals made by the artillery projectiles bursting in the air and the radio on the ground told the German aviators that our bombardment squadron had no pursuit protection and was an easy victim. The attacks of the German pursuit ships were carried on, up to within 50 feet of the bombardment planes. The next airplane to be hit was No. 13; the 2-seater caught fire and dropped its movable gasoline tank. It dived at a sharp angle, turned over on its back about 200 meters below the squadron, lost its left wing and then crashed to the ground. At this same moment a German pursuit ship was shot down on fire. No. 2 bombardment airplane was hit in the gasoline tank in the upper wing and caught fire, but the machine flaming like a torch kept its position in the formation. The machine gunner was magnificent in his courage, fighting the hostile airplanes while the flames slowly crept around him. The plane continued to fly for about 200 meters, leaving behind it a trail of fire about twice as long as the ship itself. Pilot and observer by this time were consumed and the airplane dived to its doom. At about that time a German Fokker plane diving vertically with its engine full on, lost both its wings. Now the whole right wing of the squadron had been shot down and a rearrangement of formation was made so as to again get the remaining machines into a V formation. Machines Nos. 9 and 14 were then both hit at the same time, No. 14 catching fire. The pilot of No. 14 stretched out his arms toward the sky, and waving his hand and saying farewell to the remainder of the squadron, went to eternity. No. 9 machine disappeared, and as it did so an additional German pursuit machine retired from the combat crippled. No. 15 machine was now having a hard time keeping up with the formation. Its gasoline tank had been perforated by bullets, its aileron control cut, and its rudder hit. However, it kept up.

By this time the squadron had come back to our lines, and was joined and protected by our pursuit aviation. The combat in its intensity lasted for 40 minutes, and of eighteen airplanes which had constituted the squadron, only five remained. Most of the crews were wounded and their planes perforated in all parts by bullets. They had never broken their formation, nor failed to obey the orders of their leader once. They furnished an example of military precision and bravery which is required of all airmen. These pilots were not equipped with parachutes which might have saved many a good man, nor did they have radio with which to call for assistance to their comrades not far off. In those days, we did not have such things. Today it is inexcusable to send men into combat without these two great protective measures—the parachute and the radio.

I mention this battle to show the character of training that is necessary for our people in the air. Combats of this kind during intense operations are a daily occurrence. Where once control of the air has been established the effect is terrific. It is cumulative and constantly becomes greater. It is a very different matter from anything on the ground or on the water.

Our third great branch of aviation is what we call attack. It is designed to act close to the ground and to destroy ships on the seas or on canals, railroad trains, motors, convoys or anything of that nature. It attacks from two or three hundred feet altitude and utilizes features of the ground, forests, hills, valleys or anything of that kind to conceal its movement.

Training men for air operations requires a long process and as our losses are very great among the pilots, a large number of replacements always have to be provided for. In the air when we lose a man, he is gone for good. Very few of our pilots are wounded; they are killed. On the ground, the percentage of losses is only about one killed to eight or ten wounded. A large proportion of these recover and can rejoin their units at a later time. Our airmen can never rejoin.

During the heavy air fighting in the European war, we had to figure on completely replacing the personnel of an active squadron about once a month, that is, killed, wounded and missing sometimes ran as high as one hundred per cent.

The Americans make remarkable pilots. To begin with, we have such a large number of suitable young men to draw from. Our young fellows who attend the universities, engaging in football and baseball, polo and other games of that kind, make remarkable aviators. Europeans thought that it would be difficult to discipline these young men but their training in the games mentioned above gave them an idea of team work and cooperation which was superior to any other discipline that could be instilled into them. It took no time at all to give them the idea of initiative, to support their comrades, and to evince courage in the face of danger. Few countries are able to turn out men of this kind; actually, only about five countries can turn out suitable fighting air personnel that can stand losses in the air.

When a country's pilots cannot stand heavy losses in the air, it can never form a military air force no matter how good they may be in time of peace or away from an enemy. To select these young men requires great care. They have to be physically examined, to see that their eyes, lungs and hearts are perfect, and their sense of balance is of the best, and that they are of proper temperament for the work. Of the physically sound officers accepted into an army less than 25% are suitable for aviation from a physical standpoint. In the graduation class at West Point of 1923, only about 27% of the young men were fit physically to go into the air service. The young men that enter the military academy at West Point have to undergo what is considered by the ground army to be very severe physical tests.

The young men for the air force, having been selected from those examined, are put in aviation schools where their flying and mechanical instruction is begun. They are first taught the parts of the machine, its engine and its controls. They are taken into the air daily by an instructor, taught the feel of the machine, to make turns, how to fly on the level and how to prevent going into a stall. A stall is the most dangerous of all things in the air. This means that the machine loses sufficient speed to sustain itself. An airplane requires a certain speed to be able to hold itself up in the air. One might say that it has to cover a certain number of molecules of air in a certain time to hold itself up. If it does not cover these in a certain time, the power of the air to sustain it will not be sufficient and it will fall. The most difficult thing to teach pilots is to have them feel the exact time when the stall is coming and incline the airplane downward so as to keep up sufficient speed for its sustentation. The next thing is to teach them to approach the ground at a reduced speed and to level the airplane at sufficient heights from the ground so that all the excess speed will be taken out of it before it lands. This levelling off, as it is called, at the proper altitude is also a difficult thing to teach the young flying officer. Next is taking the machine off the ground, that is, holding it down until it gets sufficient speed for its sustentation, then gradually allowing it to climb. With all of this training, if the slightest mistake is made, two or three hundred feet from the ground, death usually is the result; because at two or three hundred feet a mistake cannot be overcome. It takes this distance for a machine which is going into a stall to recover sufficient speed to fly. At greater altitudes an experienced pilot may stall his airplane with impunity, let it fall, recover speed and go ahead.

From three to four months are required before the student aviator should be allowed to fly alone. The first solo flight is the great event in a pilot's life. When he is removed from his instructor and takes the machine out alone, no matter how expert he may be, the feeling, that everything entirely depends on him with a new and strange apparatus in a medium to which he has never been used to doing things in before, is appalling. He soon gets over it, however, and then begins flying across country, using his maps and landing in strange places. All of these require different arrangements for landing and taking off. He is no longer on the nice flat airdrome but has to land in woods, fields surrounded by trees, on the side of a mountain interpersed with large rocks, or on a narrow strip of land by the sea or roadway with marshes on either side.

He is taught to fly through rain and storms, to

maintain his direction in the clouds and to fly by night. When his primary flying instruction has been finished, the pilot's record indicates what special branch of aviation is suitable for him to take up.

He then goes either to the pursuit school, the bombardment school or the attack school which specialize in these particular branches of aviation. There he learns all the ins and outs of the equipment, how to manipulate his weapons, how to fly in formation in his squadron, how to handle the mechanic's outfit for the upkeep of his airplane, what to do with his ship in case of battles in the air, in case of attack from the ground or in case of attack against the ground. After about a year's training the pilot is capable of joining the regular organization, capable of doing duty against an enemy or of working in the open country in the prosecution of any of the tasks assigned to aviation. An additional year, however, is required before a pilot really becomes the expert he should be. Many never attain this end.

I believe that it takes at least two years to make a suitable flying officer. You may teach a man to stagger around in the air in about three months. You may teach him to specialize in a given branch if he is an apt pupil in four or five months more; but to teach him every trick of the trade, to have confidence under all conditions, in the mountains, on the plains, over forests or over water, takes at least two years.

To send a pilot trained according to the principles that were used during the war against an expert pilot of the present day would mean his being completely lost. If he engaged in pursuit combat, one or two moves to the right or left, his turn, his climb or dive would be noted by his opponent at once, just in the way that an expert boxer may show up the vulnerable place in his opponent, in a couple of passes. Instantly the expert would take advantage of the weakness of his opponent and shoot him down. It is sure death for a poorly trained pilot to accept combat with an expert.

The pilot is not the only part of the personnel of aviation that needs high specialization. The mechanics that keep the airplane in the air in their way are as important as the pilot. An air mechanic is entirely different from any kind of a soldier or sailor. He is a specialist in the mechanics of an airplane, the engine, the rigging of the airplane, the upkeep of its armaments and guns, its radio telegraphy, its photographic apparatus, the oxygen for use in high altitudes and of the instruments that have to be used in navigation. These experts should form about one-half of the total enlisted part of an air force. To attempt to use soldiers or sailors as mechanics merely jeopardizes the lives of the people who go in the air.

On my first visit to our air troops along the border after my return from Europe in 1919, I found cowboys who had scarcely taken their spurs off attempting to keep the Liberty motors in order. Imagine what a proceeding of this kind means to the pilots who have to fly the machines.

The nations that have not adopted separate services,

distinct from the army or navy for their air forces, still labor under the terrible handicap of inefficient air mechanics. This endangers the lives of the pilots and makes them lack confidence in their equipment with the result that good work cannot be done in the tasks assigned them.

Mechanics in an air force should receive a compensation equal to what expert mechanics get in civil life and should be housed and cared for on a scale which is the equivalent of the living conditions of the expert mechanics in civil life. These men must be handled as expert mechanics, not as infantry soldiers or seamen. To see these expert men taken out and drilled in close order infantry formations for hours a day by order of some ground army officer is disheartening. They should do only sufficient exercises to keep them in good physical condition so that they may work on the airplanes. Taking up hours of the time in so-called ground military instruction and duty when they should be working on their airplanes is a very distressing sight.

Besides the mechanics, certain men are necessary to chop wood, haul water, clean the airplanes, do guard duty and the heavy work which requires comparatively little brains but strong backs and much muscle. In the complete creation of a military air force the following numbers of men are required in proportion to the number of machines:

For every airplane serving with an active organization on the front, there should be three airplanes in reserve, one in the interior of the country, one on the way to the front and one immediately behind the front. In the interior, where the men are being trained there should be one airplane for training to every one on the front, and one in reserve for every one on the front, so that for every airplane on the front there should be five in addition. For every airplane on the front there should be three pilots and two student pilots, one with the airplane itself, one in the first echelon of supply immediately behind the front, one on the way to the front and two in the schools. The number of observers and machine gunners for 2-seater or larger airplanes should be one-half the number of pilots. The pilots and observers are known as navigating personnel. All non-navigating officers should be composed as far as practicable of those who have already been pilots or observers. Exception should be made in the case of specially qualified engineers and specialists who should be brought into an air force on a staff basis but given no command of fighting units which they, of course, cannot be expected to control. They should exercise jurisdiction only in their particular specialties. All the navigating personnel should be officers. Non-commissioned officers and mechanics should be given the opportunity to perfect themselves to a sufficient extent to become officers. Of the men on the ground that make up the air force 50% should be expert mechanics and 50% should be of a class that is suitable for doing ordinary duties of guard, police and upkeep of an airdrome. For every airplane on the line there should be twenty men in the air force, ten of whom should be mechanics and ten unskilled men. This looks as if it would be necessary to maintain a great many men constantly with an organization to insure its efficiency and upkeep. Such is most decidedly not the case, because the men of an air force could be kept in an advanced degree of instruction while occupied with their civil pursuits. Mechanics working in automobile factories would require very little instruction to make them expert air mechanics; those in woodworking trades, metal trades, and those engaged in the manufacture of telegraphic instruments and equipment are equally proficient for air work. As a matter of fact, it is better for them to be working in civil life most of the time than to be kept in garrisons like troops in an army or seamen in the navy, because these rapidly get into the ruts of routine existence, they are apt to lose their keenness and their capacity for work. The total active air force required to defend our republic now is about twenty-four hundred combat airplanes. The distribution of the air force should be made in accordance with the density of population in the whole country. Every city of five thousand inhabitants or over should be assigned its proportion of air units. These could be stationed close at hand and the material stored in depots. Only about 10% of the personnel need be actually on duty with the units; all the rest could be with their reserve.

Certain air organizations of course are necessary for the defense of our frontiers and our outlying positions. These should be kept up to full strength as they might be needed any instant and could not be supplied conveniently in some instances in case of trouble. The airdromes should be easily reached from the centers of population of the cities. They could be used for commercial and civil purposes as well as for military purposes. The airways connecting them should be under one general government jurisdiction. The pilots and observers should be required to fly or to be in the air about one hour a week or four hours a month and should be turned out for maneuvers for from three weeks to a month a year. During the time they are not turned out for maneuvers, if the pilots are engaged in civil aviation or commercial aviation or in any kind of flying, they should be given credit for this in the same way that they would be if they were flying in the government machines. A system of this kind would be simple, economical and efficient, and would draw the best young men in the country into it, both as pilots and observers, mechanics and ground men. To attempt to handle or run an air force on the lines that an army or navy is run is to curtail its efficiency in every stage of an aeronautical undertaking. Until we have definite laws on which to organize an air force—laws in accordance with those indicated above—we will have no true air power developed.

## IX

## THE OBTAINING OF THE AIRCRAFT AND EQUIPMENT FOR THE FLYERS

THE second great requirement in the organizing of air power is the creation of suitable aircraft and equipment for the men that have to fly them. These must be devised, tried, experimented with and manufactured in an efficient manner. A true solution of the problem of national defense must be arrived at in order that suitable aircraft may be built, because the kind of aircraft necessary depends on what they have to do. To follow blindly what another nation does is merely to invite disaster, because every nation has its own particular problems to handle. An insular country, close to a continent, is at a great disadvantage when the centers of population and supply of the continental power are far away from the insular power but when the actual coast of the continental country is close to the insular power. The air units of the continental state can push right up to the coasts and attack the centers of power of the insular country in a few minutes, whereas the centers of power of the continental countries may be distant several hours from the insular country. The principles of the use of aircraft, their characteristics and their strategical handling must be decided upon entirely according to the situation the country finds itself in. The insular country in this case would have to organize its pursuit aviation primarily for the defensive, because to take the offensive at a distance and put down the hostile air force might require an air force of such size that it would be impossible to build or maintain in case of war. The effect which air power can have on an adversary diminishes as the distance to the point of attack increases.

A pursuit airplane organization for the defensive requires great climbing power, very rapid maneuvering ability, and a large ammunition supply for the planes. It may sacrifice speed and radius of action to make up for these things because it is operating over its own country and can remain there and fight—"dog fight," as we call it. On the other hand, pursuit airplanes to be used at a distance from the home land need great power of offense, speed to bring an adversary to battle, great diving capabilities for attack and a comparatively large gas capacity to go long distances. In addition, a certain proportion of pursuit aviation will be required to combat hostile airplanes trying to break through at very high altitudes, anywhere from 25 thousand to 35 thousand feet. These require special equipment for maintaining the power of the engine at those altitudes. This is obtained by using what is known as a super charger. The ordinary super charger is a turbine which is actuated by the exhaust from the engine. This in turn drives an air pump which compresses air and transmits it to the carbureter, thereby keeping the volume of oxygen constant at the very high altitudes and similiar to that found in the air at sea level. What makes an engine fall off in power at high altitudes is that the gasoline does not get enough oxygen to combine with it in the rarified atmosphere found as one ascends, and to keep up power oxygen has to be supplied artificially. At these altitudes, also, oxygen has to be furnished the pilot, observers and crew. Airplanes have to be constructed specially in accordance with each need.

With the present stage of development, our country needs pursuit airplanes of all three categories. First, those organized for the defense of large centers of power, such as the City of New York, the Pittsburg iron district, the Panama Canal or places of that kind. These should be airplanes capable of very great and rapid climb, of great maneuvering ability and large ammunition capacity. Speed and duration in the air may be sacrificed to these.

For offensive aviation designed to keep all aircraft away from our borders, to attack any vessels at sea, to be able to fly to Europe or Asia in case of necessity, we should have airplanes with a cruising ability of about eight hundred miles. About half of their fuel should be in removable gas tanks. In case combat is engaged in, the removable tank could be dropped which would lighten them and give more speed and maneuverability. The tank itself could be equipped with a fuse and when dropped act as an incendiary bomb.

These planes should have a speed of two hundred miles an hour, have several guns, at least one of which should be fifty caliber. About twenty per cent of these pursuit airplanes should be constructed for high altitude work and be capable of going up to thirty-five or forty thousand feet. With a pursuit aviation of this kind our bombardment aviation could be accompanied and protected on its missions whenever necessary. Seacraft or hostile aircraft could be attacked three hundred miles off the coast, a trip across the continent could be made which should involve only three or four stops, and a range of speed of from 135 to 150 miles an hour could be maintained. A pursuit force of aviation of this kind could protect the whole triangle, Chicago, Bangor, Maine, to the Chesapeake Bay from a central point in the northern part of New Jersey, within three to four hours. The triangle mentioned above is the strategic heart of the United States.

In the types of airplanes to be produced, constant improvements and a thorough knowledge of the possible adversaries' equipment are necessary at all times. We must excel, and this excelling must be a continuing thing. Once a nation has dropped behind in its development, it is like making a stern chase, and a very difficult undertaking again to get the lead. It is entirely impossible of accomplishment without great knowledge and ability at the top.

The general development of aircraft should be patterned on the following lines. First, there should be airplanes in the hands of the air troops that are better than any in existence. Next, there should be airplanes being made in the factories to take their places that are better than those in the hands of the troops, and third, there should be designing and experimental work being done on airplanes that are better than the other two. An airplane in being with the squadrons; an airplane in production in the factories, and an airplane on the drafting table waiting to be experimented with—this system is being followed now by all the great air powers.

A system of this kind has to be thought out years ahead of time. It cannot be adopted on the spur of the moment, and if a continuing policy of aircraft construction is not adopted an air force cannot keep up to date with safe equipment in the hands of its pilots. Many instances to illustrate this have occurred in this country. In 1919 we devised a super-bomber, capable of going 1300 miles without landing, to carry two four thousand pound bombs, and to be able to land or hit the ground at a speed of one hundred miles an hour without smashing up. It was equipped with six engines with the idea of being able to fly on half of them and was otherwise provided with all sorts of navigation equipment so as to be able to fly in fogs and at night. The first one of these planes was completed and was entirely successful from an experimental standpoint. It was built according to the old system, had three wings of trussed construction and many other old features. The knowledge that we gained from this airplane should have immediately been put into another and the following airplane should have been a large monoplane with the engines put in the leading edge of the wing and the landing gear made of wheels, ten or fifteen feet high, with skis to land on the snow and the body of the airplane made water-tight so that it could float on the water. This was never constructed. From this type of airplane we should have built a third which should have been completed this year, 1925. It would have been an all-metal construction monoplane, designed with from four to six engines, driving through gears onto one large propeller, giving a speed of 135 miles per hour. It would have been capable of landing on the earth, on the snow or ice, or of floating on water. This air-plane would have had a 2500 mile radius, would have been able to fly from New York to Pekin in three jumps and could have gone at this time to rescue the Amundsen Polar expedition or to Australia or Asia via South America. A lack of development of this kind of a plane has put us behind so far that we cannot catch up without going to Europe now for a model for it. If we had continued this development we would have led the world in large airplanes.

The bombardment type of airplanes should be cap-able of carrying two projectiles, either of which is capable of crippling, destroying and sinking the largest battleship afloat, or of carying one bomb which is twice as powerful as either one of these. The reason for this is that an airplane may miss its first shot and be in a position to correct its aim and make a hit with the second shot. This gives much more of a factor of sureness of a hit because the probability of hitting with the second shot is many times as great as hitting with the first shot; also, an airplane, capable of carrying two of the largest projectiles can carry one in the shape of a bomb—a two thousand pound bomb, for instance, and another in the shape of a water torpedo or aerial torpedo of the same weight. At the present time, two two-thousand pound bombs or one four-thousand pound bomb is necessary for our bombardment aircraft. There need be only one general type of bombardment airplane. It should have a speed of about 135 miles an hour and a cruising range equal to that of the pursuit aviation or about eight hundred miles while loaded with bombs. It should be equipped with super chargers so that it could act at any altitude from ground level up to thirty-five thousand feet.

If the bombs are taken off and in their place gasoline in removable tanks is put on the airplanes, their cruising ability is more than doubled. Instead of being able to go eight hundred miles they would be able to go from sixteen hundred to two thousand miles. By increasing the gas capacity in this way they could land and refuel the pursuit planes that are accompanying them so as to give them also greater radius of action. Should it be desired to use bombardment airplanes as transports they can carry tons of equipment in the form of bombs, ammunition, food, spare parts, mechanics or anything that is necessary. In other words, a properly organized air force can maintain its own communications and supply through the air to a very great extent. In the fall of 1921 when the domestic disturbances in West Virginia took on the character of an insurrection and Federal forces were sent into that area, two squadrons of my air brigade at Langley Field were ordered to Charleston, West Virginia, for duty. This place is situated in the middle of the Allegheny mountains and hard to get to on account of the lack of communications. Four hours after receiving the order the air forces were on the way. These were the same aircraft that had sunk the battleships a hundred miles at sea. The squadrons consisted of two-seater D.H. planes and were supplied with medical attendance, food, equipment, bombs and ammunition and spare parts by Martin bombing airplanes acting as transport ships. This was one of the first times that this had been done in actual service. Since then we have supplied ourselves under many conditions including work in the snow and ice during the wintertime in Northern Michigan. The proportion of bombardment aircraft to pursuit aircraft should be two pursuit airplanes, twenty per cent of which are for high altitude work, to one bombardment airplane. With heavy air units of two hundred pursuit airplanes and one hundred bombardment airplanes properly organized, equipped and manned, this country need have no fear of invasion either through the air or over the water. Without these it is helpless.

The third category of airplanes is known as the attack type. This was developed to act directly in conjunction with ground troops during the European war. It is a question whether it will have the application in the future that it had in the past because air power should be designed to strike at great distances where the enemy is concentrated, either in manufacturing districts, coming on railroad trains or steamships, or otherwise confined in a small space. This is where aviation renders its greatest service. To demoralize an enemy unprovided with aircraft or not well provided with aircraft, against savage tribes or poorly organized levies, attack aviation has a tremendous effect. These units should be organized with special reference to the particular object in view. This airplane also should have about 800 miles cruising ability and should be capable of landing and fighting on the ground with its equipment, it should have from four to six machine guns, one of which should be either a 50-caliber gun or 37 milimeter or a one-inch cannon. They should be organized into groups of one hundred airplanes each, and in accordance with the air strength of the enemy should be assigned pursuit aviation for their protection. If acting along a coast or frontier or where a certain locality has to be swept clean, they should be protected by pursuit aviation of the defensive type such as was explained above to be necessary for the protection of localities.

The reason that attack aviation should be protected by pursuit aviation of the defensive type is because in their operations they attack and re-attack their target. For instance, if they are attacking a long column of motor transport they wait until it either gets in a cut in the road, a defile of some kind, a causeway with water on each side or some other place where the motor transport cannot turn off the road.

They then attack the leading trucks and the tail-end trucks and set them on fire so as to block the roads. They then continue their attacks until the whole train is destroyed. The pursuit aviation, therefore, has to hover over the locality and engage any hostile pursuit aviation that may attempt to dive on the attack ships while engaged in destroying their targets. This system of attack of course applies to columns of troops, railroad trains, canals, ships or anything else that requires their attention.

The proportion of pursuit aviation for their protection should be about the same as for the protection of bombardment aviation; that is, two pursuit airplanes to one attack airplane.

Airplanes should be of all metal construction so that they could be stored without deterioration and stand out in all kinds of weather without shelter. In the future it will be impossible to build hangars or otherwise protect air forces from the weather. The airplanes must stand right out in it, winter and summer, in the heat and the cold, whether they are used in the tropics or the arctic. This result is not difficult of attainment if proper and continued experimentation along this line is carried out. They should be equipped so as to enable them to alight on land, ice or snow, or, if necessary, on water. They should have equipment and instruments so that they could fly both by day and by night, in clouds or through storms.

It is often said that it takes as long to make an airplane as a battleship. This is really quite true, because from the time an airplane is first conceived until the time it is actually turned out and put into the hands of the troops requires several years.

First it must be decided what the characteristics of the airplane should be, that is, what its speed shall be, how fast it must climb to the required altitude, how much weight it should carry, how far it should be able to go, the kind of surfaces or ground that it is to land on and a great many other considerations. These have to be thought out by the engineers. Some of them can be met and some have to be compromised. Small exact models of the airplane a foot or two long are then constructed and tried out in wind tunnels to see just how the air will affect that particular type of airplane. Wind tunnels are long tubes several feet in diameter. The wind in these tunnels is actuated by fans up to speeds that correspond to those which the airplane will encounter when going through the air at its designed velocity. The small model airplane is suspended on delicate balances which show every pressure which the wind exerts on it. These measurements are becoming so exact today that when the first model of the full-sized airplane is turned out, it is almost certain to fly well at once and answer the controls. Of course, there is a certain amount of the cut and try system still necessary, but construction of airplanes is becoming more of an exact science all the time. When the aerodynamic characteristics have been proved in the wind tunnel, what is called a full-sized "mockup" of the airplane is made. This is a wire and wood construction of exactly the dimensions of the full-sized plane. In it are arranged in their appropriate places all the instruments, the controls and accessories which the airplane has to have. The "mockup" then serves as a model on which the service airplane can be patterned.

After this comes the creation of what are called "production drawings." These are extremely hard to make because each airplane has hundreds, often thousands of parts. Each part has to have its drawings so that they can be made by the hundreds or thousands when required. If each airplane had to be made by hand separately, the cost would be prohibitive, and it would take so long to make them that they would be obsolete before they were ever constructed; also, there would be no interchangeability of parts.

Some factories in this country if transformed into airplane production centers could turn out single-handed three or four hundred airplanes per day if furnished with proper drawings. The drawings, having been completed, are given to the factory which has been chosen to make the airplanes and a few are produced, say eight or ten. These are turned over to the troops and the technical services where they are thoroughly tested and subjected to strains and stresses to see whether they are strong enough to meet all conditions of flying. Again, changes are required before they can be finally issued. This, of course, requires further changes in the original drawings. If these are interfered with too much and too many changes are made, of course it increases the time of delivery and makes the cost mount up very high. This is particularly injurious to a manufacturer who has made all his jigs and arrangements for producing a certain type of airplane. The government agency, therefore, ordering them has to be very careful to make their specifications, drawings and requirements very exact so that they can be made by the manufacturer as ordered.

Various ways have been tried by governments to obtain suitable aeronautical equipment. At the beginning of the art of flying there were not sufficient people in civil life to take up the work, nor was it sufficiently remunerative for them on account of the fact that there was no civil aeronautical effort. All airplane orders had to be placed by the government and these were almost entirely for military purposes. This gradually led up to the usual government monopoly of aircraft and developed into the government itself taking over all the engineering and designing of airplanes. This system stifles initiative on the part of the citizens and in many instances led to the government establishing aircraft factories which crowded out private and civil factories, increased the cost which government construction always does, and resulted very largely in holding back invention.

The leading aeronautical countries have now come to a pretty definite system in the ordering of their aircraft. It is carried out as follows: The operating division of an air force, that is, the people that have to use the airplanes, decide on what kind of machines are necessary. This data then is transmitted to what is known as the engineering division or technical section which interprets the desires of the operating men into technical language which will be understood by the aeronautical engineers. They are invited to submit their ideas and bids on how the government requirements can be met in the new airplane. The data furnished by the engineers is then considered by the engineering division and awards made to the civilians submitting the best designs. Usually a cash payment is made for the design itself so as to encourage these engineers, and a few airplanes of this type are then ordered. The production drawings are completed and again a competition takes place to determine what factory shall make the aircraft. Not all manufacturing concerns are capable of making suitable aircraft. Those that make a business of it to the exclusion of other things are encouraged above those that are not as capable of turning out suitable airplanes. It has been found that a preferred list of aircraft manufacturers who have the ability to turn out good equipment, works out to the benefit of the country. These then are given the bulk of the aircraft orders and a careful cost accounting is made by the government in order to see that excessive charges are not made against it. Should any new factories desire to start and come up to the standard of the older ones or exceed them, they in their turn, will be put on the preferred list. This system, however, prevents factories really incapable of building suitable airplanes for the flyers from horning in and getting the orders by bidding lower than older and more experienced manufacturers and in some cases getting away with it by political pressure. The greatest care has to be constantly exercised to prevent troops from getting worthless airplanes and in constantly keeping the aircraft in suitable condition for flying under all conditions. If politics get into the proceeding, it is manifested by good-for-nothing airplanes and deaths among the pilots.

The organization which has been found best to provide for the construction of aircraft is as follows: First, the engineering division of an air force interprets the desires of the operating flyers into such words that the civilian engineers of the country may be able to understand them and submit their ideas. This organization must be able also to test the strength of the materials and the flying ability of the aircraft. Second, the production organization gives out the orders in a suitable way to the civilian manufacturers for the production of the airplanes. This organization should provide for a proper inspection of every part and piece of the airplanes according to a system prescribed by the engineering division. Third, the new aircraft should be concentrated in the aeronautical supply depots, assembled and always kept in condition for issue to the air troops.

No aircraft should ever be manufactured by the government on a production basis. This always results in inferior equipment, excessive cost and even worse political control than was found to be the case where private manufacturers were employed.

The supply system of an air force should consist of three general divisions. The main supply depots are in the home country, where the material is brought together from the manufactories and assembled preparatory to sending it to the troops. Those charged with the issue of aeronautical equipment of all kinds, the airplane itself, the guns and cannon, signalling devices, radio equipment, cameras, instruments of all kinds, motor transportation and all accessories require a thoroughly efficient, carefully trained organization.

Each supply depot should be organized into three principal departments. One, a supply department which issues all the material. Second, the repair department which repairs any of the equipment requiring renewal or replacement; and third, the salvage section which recovers and goes over all broken or damaged airplanes or material which are shipped into the depot, saving all the useful parts which can be used over again and getting rid of the others wherever they can be sold or profitably disposed of. A good salvage department results in the saving of a great deal of money. The main supply depots should hold usually a year's equipment for any likely undertaking, because ordinarily it will take that length of time in case of war to organize properly the manufacturing facilities of a country to supply it with aircraft. There should be another supply organization immediately accessible to the troops and close to them, which can be moved from place to place and which should be constantly supplied by the main supply depots. These depots were designated by us in Europe as air depots. They always had 100% of airplanes available to send to the troops, and as fights took place the airplanes used up by the air units were immediately replaced from the air depots. In addition to the airplanes, these air depots carried a two weeks' supply of fuel, ammunition and all accessories, and in their turn were organized into supply, repair and salvage sections.

The last supply elements are the air parks that are maintained right up with the air troops. These should carry about two or three days' supply constantly on hand, and be capable of making repairs of about the same magnitude that a chauffeur would be capable of making on an automobile with tools that he has on the car. These parks should also be organized into supply, repair and salvage sections. They should handle all supplies for the operating units to which they are attached. The same general system of administration, supply and engineering direction should apply to all three echelons in the supply system: The main supply depots, the air depots and the parks.

Of course, to make for efficiency it is necessary that we have just as few types of airplanes as possible. At the time of the Armistice in Europe we had eleven distinct types of airplanes on the front; probably one-half of this number would have done. If we had been able to organize our system as we desired to have it, and as we recommended at the beginning of the war, a saving in the number of men used in the supply department and the transportation required for spare parts both for planes and motors would have reduced the effort necessary more than one-half. An airplane has to be supplied constantly with spare parts. After every trip, almost, something has to be replaced, and whenever an airplane is ordered it should be ordered with the requisite number of spare parts to provide for its flying life; otherwise another airplane has to be torn to pieces to get spare parts for the one being repaired.

Constant development and experimentation must go on to keep up with the nations most rapidly gaining in the art and science of flying. We must project our vision at least seven years ahead of time to see what is necessary. If sufficient aircraft are not on hand at the beginning of a war, supremacy in the air can never be obtained because the whole country will be subject to air attack and the means of manufacturing airplanes or providing a trained personnel will be either completely stopped, or inexperienced men will have to take to the air with inferior equipment to offer battle to an experienced enemy with entirely up-to-date equipment. If we had not been supplied with the best equipment made by the Europeans in the last war we would have stood no chance, because America had failed entirely to develop any air power worthy of the name. It must not be thought that because the United States has capabilities of manufacturing aircraft more quickly and in greater numbers and more cheaply than other countries, that we should wait until another contest has started, to build aircraft. This is a most decidedly mistaken policy and should not be entered into. Without our own means of supplying ourselves with modern aircraft both in quality and in quantity, we cannot, as a nation, take any independent action among the great powers of the world, because we always will be subject to their control as we were in the last war.

# X

## THE DEFENSE AGAINST AIRCRAFT

It was proved in the European war that the only effective defense against aerial attack is to whip the enemy's air forces in air battles. In other words, seizing the initiative, forcing the enemy to the defensive in his own territory, attacking his most important ground positions, menacing his airplanes on the ground, in the hangars, on the airdromes and in the factories so that he will be forced to take the air and defend them. To sit down on one's own territory and wait for the other fellow to come, is to be whipped before an operation has even commenced.

During the Chateau-Thierry operations when our air forces first came up on the line in the last part of June and the first part of July, 1918, the Germans controlled the air completely. They had concentrated the major part of their aviation against the air force assigned to the Sixth French Army and destroyed it, while their offense on the ground was so rapid that they captured the French airdromes with the planes still in them. We were ordered from our concentration area in the Toul region to reinforce the Sixth French Army that was holding the line of the Marne River. We started there on the 28th day of June,

1918, arriving that night with all our planes. We were under French command and they assigned us to what is known as barrier patrol duty along the front. This consists of having a flight of five or six airplanes or maybe a squadron patrol back and forth on a front of about ten miles, which makes both ends of such a patrol sector visible from the other end. The barrier patrol was designed to keep out any hostile aircraft that might come into the area. These patrol areas were joined on each side by other patrols, and in this way the whole front was covered. If only one or two German planes attempted to break through, this system worked, but it was fallacious because the Germans, finding out the strength of the patrols, could concentrate their aviation and outnumber any one of our patrols three or four to one, jump on it, destroy it and go ahead and do whatever they wanted to.

In a few days our losses were terrific. Among the lost were Quentin Roosevelt, Allan Winslow and other valuable men. In the meantime, some of us had flown clear across the German position all the way from La Ferte Sous Jouarre across Fere-en-Tardenois to Soissons, cross cutting the whole German area. I did it alone in a Newport single seater, and later Major Brereton with Captain Hazlett as observer did the same thing. We had definitely located the German center of supply at Fere-en-Tardenois. The woods were full of ammunition, machine guns, cannon, fuel, gasoline and oil, motor transport, pontoons, temporary railroads, and everything that goes to make up an army's equipment.

We had found a place that would have to be protected in case it was menaced from the air, because a few bombs dropped on the ammunition dumps would blow them up, gasoline and oil would be set on fire, and the losses of material might be so great as to completely stop a forward movement. I therefore asked that bombardment aviation be called for immediately as we had none. The French air division, used as the great mass of maneuver of air operations, was attempting to cover the front of the Fourth French Army which was menaced by a strong German attack. They had been badly depleted and were very tired as they had been fighting constantly since the German attack against the Fifth British Army under General Goff, in March. They could not come to assist us. We therefore asked the British to send a brigade of their air force to help us. They despatched it immediately. It consisted of three squadrons of two-seater D.H. 9 bombardment airplanes, and two squadrons of pursuit, one of which was equipped with Sopwith Camels and the other with S. E. 5's. They arrived full of fight and ready to go.

On the following morning the British bombardment force was given directions to attack Fere-en-Tardenois at dawn. We concentrated all our pursuit aviation, four squadrons of ours and two of the British and converged on Fere-en-Tardenois from three different directions so as to arrive there at the same time that the bombardment did. About thirty-six British bombardment ships made the attack at an altitude of from five hundred to one thousand feet. It was a remarkably brave feat. They blew up several ammunition dumps and caused consternation among the Germans. The German pursuit ships, concentrated to defend Fere-en-Tardenois, shot down twelve British bombardment planes on that day. Our pursuit aviation, on the other hand, shot down more than this number of Germans with small loss to our pursuit. The tables had been completely turned on the Germans and they had to stand with their aviation and defend Fere-en-Tardenois. They no longer could keep penetrating our lines because if they did they would so weaken themselves around Fere-en-Tardenois as to become incapable of defending it. If they did not keep a large part of their aviation constantly in the air, on account of the short distance from our lines to Fere-en-Tardenois they would be too late to attack us; because, if they awaited us on the ground, they would be unable to climb up to us before we were well on our way back, after having dropped our bombs. Therefore, they could never oppose us with more than one-quarter of their total aviation, usually with not more than one-sixth, because the part that was not in the air when we arrived there was useless. If they put all their aviation in the air at once we would watch it and attack when they returned to the ground for fuel. We had seized the power of initiative by finding a spot very vulnerable to the Germans which put them on the defensive although they greatly outnumbered us. We concentrated our whole aviation and launched it in one body. Of course, the Germans might have done the same thing to us if we had had a place behind our lines that was as important to us as was Fere-en-Tardenois to them. There was no such place, because our troops were being supplied on converging lines, while theirs were being supplied on diverging lines from Fere-en-Tardenois. The same condition lasted practically during the rest of the Chateau-Thierry operations, because the area around Fere-en-Tardenois had to be kept full of supplies for the German army.

When seizing the initiative and carrying the air war into the enemy's country is not possible or practicable, the only other method of defense against aircraft is the use of guns and cannon from the ground, combined with the action of defensive pursuit aviation. The air is so vast and extends so far that the shooting of airplanes out of the sky, with cannon from the ground, is almost impossible of achievement, especially when the planes are almost always protected by clouds, by the glaring sun or by darkness. Once airplanes have beaten the hostile aircraft in air battles, nothing can stop their operations.

To begin with, the airplanes have to be seen or heard. Clouds, the night, storms and the sun itself conceal airplanes from view. Nowadays, large propellers are being used which make little noise and the engine itself can be muffled as well as that on an automobile, so that it is becoming more difficult to hear a flying airplane. This was not the case in the war when the gearing of the engine to slow down the total number of the propeller's revolutions was not practicable because the gears were not sure enough in their action. Consequently, the small propellers, in whirling through the air, made as much if not more noise than the engines themselves. This is the reason noiseless airplanes were not used during the European war. Aircraft then could be located, at times easily, by sound ranging systems on the ground. Now it would be very difficult.

Before the true attack on a city or locality develops, many feints are made. Airplanes are sent from various directions to draw out the enemy's means of defense. Thousands of rounds of shells from the anti-aircraft guns are wasted in the air by the defenders. The gun crews kept up at night constantly firing are soon worn out; their eyes become blinded by the incessant flashes. The whole arrangement of ground protection against aircraft, sound ranging, searchlights and guns cannot stand up under intelligent air attack and is incapable of serious effect on airplanes.

In addition to anti-aircraft guns, a network of wires was raised by balloons around important places in the World War so as to form an entanglement into which airplanes would fly. The wires were intended to cut off wings, damage propellers and otherwise bring airplanes down. Airplanes forthwith were equipped with steel-cutting wires projecting in front of them which would cut through any of the barrages. The airplanes also found the supporting balloons and shot them down, thereby completely nullifying the wire barrages. I never saw an airplane brought down by this device.

In a modern war, bombardment aircraft would be equipped with the aerial torpedo and with gliding bombs. With these they can stand off for many miles and hit a target as large as a city practically every time. A gliding bomb depends on gravity for its impelling power. They have been constructed so that they will go about a mile for every one thousand feet of altitude. An airplane 10,000 feet up could project one of these ten miles. They can be directed toward their objective, by gyroscopic control, or by radio telegraphy. The aerial torpedo of course will go as far as its charge of gasoline will carry it. It can also be directed and controlled by radio.

During the war, our means of navigation were so crude that often at night we would have to follow a coast line, a white road or river to attain our objective. Consequently, the enemy, knowing the routes the aircraft would have to follow to attain their objective, would spread out listening posts all along them, arrange their wire barrages, searchlights, anti-aircraft cannon and machine guns accordingly. Our heavy bombardment airplanes during the war were unable to rise more than a few hundred feet from the ground, which, combined with the noise that the airplanes caused then, made spotting them comparatively easy. Now, either by directional radio or by an instrument which will show how far we have gone in the direction we have taken and our altitude, we can always estimate our position very accurately, and find the place we desire to bomb without following ground indications.

Any system of defense against aircraft from the ground alone is fallacious and money put into it, if not spent along carefully considered lines, is merely thrown away.

Ground armies, unfamiliar with the action of air power, are constantly setting up the claim that anti-aircraft artillery is capable of warding off air attack. This is absolutely not in keeping with the facts, and a doctrine of this kind is a dangerous thing to be propagated, because it inclines people to think that they have security from this source whereas they have not. It is, of course, more difficult to protect a warship at sea with anti-aircraft artillery than it is to protect localities on land because the vessel is a movable platform.

Aside from the method of attacking an enemy's air force when it is at a distance, which is the only real means of keeping aircraft away, the next best way of meeting an attack is to organize a given locality such as New York, for instance, with the following system. . . . The average bombardment formation of about a hundred airplanes will move at an altitude of about 15,000 feet and at a speed of from a hundred to a hundred and thirty-five miles an hour. It will keep together until it approaches its objective so that the leader may be able to maintain control of the individual airplanes, to concentrate the attack and to allow them to return to their initial airdromes so that they can be quickly refueled, re-ammunitioned and continue the attack. If hostile aircraft are enabled by the lack of offensive action at a distance to come directly over a city and engage in combat with our own aircraft, it is manifest that, if anti-aircraft artillery fires into the air, it is just as apt to hit its own aircraft as that of the enemy, as they will be close together in combat. Many of these actions will take place at night, in the clouds or in the direction of the sun.

It takes our own defending pursuit aviation at least twenty minutes to take to the air and rise to 15,000 feet after having received the order. Usually we allow half an hour. They should be up there a few minutes before the hostile formation arrives so as to be entirely ready to launch a concentrated attack against the enemy. This requires that the hostile formation be picked up at least a hundred miles away so as to enable our own people to make up their minds as to the strength, disposition, probable intentions and number of the enemy in order to take proper counter measures. It cannot be told with certainty from what direction the opposing aircraft are coming; therefore it is necessary to completely surround the locality to be defended with observation and listening posts both in the air and on the ground.

Listening posts are not enough, because modern airplanes do not necessarily make a noise; therefore, surveillance aviation has to be kept constantly in the air where they can not only see but hear the enemy aircraft.

When we had organized our defense against aircraft to resist night attacks of the German bombardment aviation in the Argonne, in the latter part of October and the first part of November, 1918, we had listening posts along the front lines held by the troops. These were connected by wire and radio with the anti-aircraft positions which were the next echelon behind the listening posts. Behind these was the line of searchlights and behind the lights a constant patrol of pursuit aviation with areas of about five or six miles of front allotted to each airplane. Our airplanes would rise at night to about eight or ten thousand feet, would shut off their motors and glide, looking constantly for any signal lights on the ground which reported the presence of enemy aircraft, or firing of the anti-aircraft artillery. They would look and listen. When the enemy aircraft approached our front the listening posts would report them and the defense commander would try to follow them on the track system at headquarters. Information about them would be sent to all our posts, a barrage fire of anti-aircraft artillery would be started and movable searchlights would streak the sky. It was a picturesque and invigorating sight but not particularly dangerous to the attacking planes.

The theory was that if a hostile airplane came over a lighted area, all the searchlights would be turned on so as to illuminate the airplane. At this time, also, the pursuit airplane, drawn toward the enemy bomber by the bursting anti-aircraft shells and then illuminated by the searchlights, would dive at it with its engine full on and attempt to shoot it down. During the first night that we had this system in effect in the Argonne we had five combats—night aerial combats—and drove the enemy aircraft back. Never before had we been successful in doing this with searchlights and anti-aircraft artillery fire alone. The Germans, however, at this time, were becoming pretty weak in their bombardment aviation. We would attack all their airdromes every night to such an extent that they could not stay in the same airdrome two nights in succession and had to move from place to place. They would fly off from one airdrome and land on another on account of fear of attack, so that their crashes and losses were great and their system of command was much impaired. As a matter of fact, of the airplanes that were surrendered to me on our front after the armistice, there were only eight serviceable German bombardment planes.

In the case of cities on the sea coast, these anti-aircraft arrangements must be pushed out to sea. The aquatic listening and observation posts may be in the form of submarines, surface vessels or light ships, and stationed over one hundred miles away.

The complete air defense of a locality requires: First, that there be a circle of listening and reporting posts, extending for at least a hundred and fifty miles out from the area to be defended. These should be supplemented by aerial observation posts and surveillance aircraft. Second, there should be an organization of pursuit aircraft, the type that rises rapidly and maneuvers easily. Third, there should be several circles of searchlights in groups of forty or fifty each. In Europe I always liked groups of thirty lights, twenty of which were fixed in position so as to illuminate an area in the sky and ten of which were movable so as to try to pick up the opposing aircraft and follow them. The fixed lighted area was always that in which the maximum danger was to be expected. These would be flashed on and off as the hostile aircraft came over the area to be defended while the movable lights searched for and tried to follow the planes. Fourth, there should be anti-aircraft guns and cannon. All of these elements should be under one control, that of the air commander or the one charged with the whole of the air defense of the locality. He should have a large control board arranged like a map showing each part of the defense area in detail, and the whole marked in squares covering a distance on the ground corresponding to five minutes' flight of an airplane. A system of electric lights should be so arranged under this map so that the direction of flight, the speed, number and type of the hostile aircraft could be projected easily on the board. The anti-aircraft defense system should have an independent and separate telephone, telegraph, radio and courier system to every element in the command so as to instantaneously communicate with them about any arrangement desired. If the air communication system is used by other troops its efficacy is lost. An instant's delay in an air operation may nullify the whole thing because planes move so quickly. By a system of this kind every element for the air defense of a locality is brought into play, and coordinated with every other part. It is an extremely complicated, expensive and difficult arrangement to install, but it was the only method possible to use during the European war to ward off air attack. It unquestionably had the effect of keeping aircraft up at a considerable altitude so as to escape the attack of pursuit airplanes. Machine guns, cannon, or any missile throwing weapons would not do this because of the ineffectiveness of anti-aircraft fire, and because sometimes it is easier to fly down close to the ground to avoid fire from the ground than it is to keep up high in the air. The idea of being able to defend any locality whatever against airplanes with anti-aircraft guns, cannon or any other arrangements from the ground alone is absolutely incapable of accomplishment.

During the war anti-aircraft artillery brought down about one-tenth of one per cent of our airplanes that were fired at. The great difficulty about hitting anything in the air is that there is no point of reference up there, no tree or church steeple, road crossing or hill on which fire may be adjusted. It is difficult enough for field artillery to hit comparatively small targets on the ground at any range over five or six thousand yards where there is an opportunity to range and to correctly observe the fire. At five thousand yards in the air the largest airplane is a very small dot and can be seen during the clearest weather only with difficulty. Even if the altitude of the airplane is known it is very unlikely that it can be hit by shell fire from anti-aircraft guns because the fuse setting has to be so carefully done that a small change of a fraction of a second in the timing of the fuse will carry the projectile several hundred feet to one side or the other. As a matter of fact, the exact distance and speed of a plane are practically unknown to the artilleryman, and it is extremely difficult for him to determine these factors. Even if these could be found out accurately, the fuse setting of the projectile, the traversing of the gun and the aim, have to take place instantly, which is a very difficult thing to accomplish in the time given. The slowest service plane moves at about one hundred miles per hour or 147 feet a second, or in one minute and 43 seconds it would cover 5,000 yards or nearly three miles. As direct hits from single cannon against an individual airplane are next to impossible, the other hope of the artilleryman has always been to place an aerial barrage of shells around the vital area. That is, to so fill the air with projectiles that no airplane could fly through it without being struck by some of them. This, of course, is a practical impossibility, in that it would require such a great number of guns and such an expenditure of ammunition that the effort and expense would be prohibitive in comparison with the object attained. Frequently during the war we would send a succession of reconnoitering planes over an enemy position on which we wished to start the guns firing. The guns would work briskly for an hour or two, filling the air with projectiles. The gun crews would then become tired, and the expenditure of ammunition so tremendous that there would be danger of wearing out the guns. We would wait for the early morning hours, then launch the main attack and without exception we always succeeded.

During the past year a great many statements have

been made that the anti-aircraft artillery is improving and that the results now would be different from those that occurred during the war. This is most decidedly not the case. The improvements made, which amount to little, are nothing in comparison with the improvements made in airplanes, in their speed, their ability to climb high in the air, to make little noise and to conceal themselves from view. Anti-aircraft positions can be attacked directly by airplanes so as to nullify their fire and may be covered by smoke clouds or gas clouds. Airplanes themselves may shoot them up with machine guns or with bombs dropped in their vicinity either by the gliding method or directly.

The average anti-aircraft gun costs anywhere from twenty to thirty thousand dollars. They will fire about twenty shots a minute with each shot costing from twenty to thirty-five dollars. The life of these guns is from about fifteen hundred to two thousand rounds when they must be replaced.

If a locality is defended from the air passively, and tied up to one place in much the way that a seacoast artillery gun is, an attacking air commander of ability, seizing the power of initiative, can arrange his air raids, make feints and confuse the enemy, so as to be able to launch his main attack practically when and where he desires without much loss to himself.

The only defense against aircraft is by hitting the enemy first, just as far away from home as possible. The idea of defending the country against air attack by machine guns or anti-aircraft cannon from the ground, is absolutely incapable of being carried out.

## XI

## CONCLUSIONS

The development of air power has forced a complete reorganization of all the arrangements for national defense.

The rapidity and sureness of electrical communication all over the world make it possible to combine the use of all the elements entering into national defense in a manner impossible of accomplishment heretofore.

The influence of air power on the ability of one nation to impress its will on another in an armed contest will be decisive. Aircraft of certain classes are now able to traverse the air all over the world no matter whether they be over the sea or over the land.

Consequent upon this, the mission of each branch of the national defense must be clearly stated and its powers and limitations thoroughly understood in order to combine its action with the other branches to insure the maximum effect.

Before the coming of air power, the national defense elements consisted of land power and sea power. At that time, all operations over the sea were assigned to the Navy. Everything on or over the land was assigned to the Army. There was a little overlapping of duties between these services immediately along the coasts, but this was not a serious proposition.

Now, however, air forces operating from land bases can control the surface of the sea and the air over it up to their operating distance from the coast. Within that distance a navy no longer has the paramount interest. Therefore the Navy's mission so far as coast defense is concerned, has ceased to exist and its mission must be beyond the zone of aircraft activity.

The land organization of a navy for coast defense and the land establishments incident to it can be dispensed with. The money and effort heretofore put into these should now be placed in air defense organization.

The mission of land power and the army will remain very much the same. The modifications necessary will consist largely in concentrating gun power around the major units of infantry and giving them their maximum mobility. Coast artillery, except at points that can be affected by gun fire from submarines, has become superfluous and the money and effort put into this should be transferred to air power.

Air power, however, must be assigned a definite mission in its particular sphere of activity. This mission should be the responsibility for the complete air defense of the nation. Without a mission of this kind being specifically prescribed, the aeronautical effort of the country will be dissipated between the land, water, and other services, so that maximum efficiency *cannot be obtained.*

All of the great countries of the world are now organizing their air power for striking their adversaries as far away from their own countries as possible, whether the enemy be in the air, on the water, or on the land. This policy is adopted so as to make the home country free from the interference of hostile forces by keeping the fighting at a distance from the frontiers or coasts.

The underlying principle in the organization of air power is the creation of an air force capable of the greatest radius of action practicable under the conditions limited by personnel, material, and armament.

Next in the air force in order of importance is the organization of certain local air units destined for the protection of extremely important centers of power. The City of New York serves as an example of a locality of this kind. Such local defense units should combine all means and methods used for defense against aircraft, both in the air and on the ground.

Third, in order of importance, are the auxiliary air units assigned to military organizations in the air, on the water, and on the land. This class of air unit, ordinarily, is called observation aviation. Like all auxiliaries, it should be cut down to the lowest point commensurate with the efficiency.

Therefore, in the organization of our air power we should consider:

1. The air force.
2. Local air defense units.
3. Auxiliary air units.

Two and Three, in addition to their other duties, should be trained to assume the offensive.

The system of command of military air power should consist in having the greatest centralization practicable. An air force now can move from one to two thousand miles within twenty-four hours. Military elements on the land or water can move only a fraction of this. If, for instance, we take an air sector from the Chesapeake Bay to and including Maine, approximately eight hundred miles in length, which can be covered by an air force in eight hours or less, several sectors of defense would have to be organized by the ground troops for this distance. To assign air force units to any one of these ground organizations would result in the piece-meal application of air power and the inability to develop the maximum force at the critical point. Therefore, all air force units should be directly under the orders of the Commander in Chief of the military power of the country.

Local air defense units destined for the protection of a locality likewise should be under the control of the supreme commander and also in close liaison with the ground troops in their vicinity.

Auxiliary air units should be under the command of the military elements to which they are assigned, subject to the general control as to training, sector, depth of reconnaissance, methods, and supply by the air force.

The establishment and control of airways and the seizure of aviation bases in offensive operations, in which the air force has a dominant interest, should be under air force control. The air force also should have the decision about the recruitment and training of all air force personnel, the procurement of all aircraft, all flying material, armament and accessories.

Air power is the most rapidly developing element in the makeup of nations. Accurate vision is required to keep abreast of the times and programs must be adopted on the basis of what will happen from seven to ten years hence. Failure to estimate properly what will occur will result in serious consequences in case of war.

Considering our possible emergencies in the future, the following Air Force should be provided for:

Within the continental limits of the United States there should be an offensive air force of one air division of twelve hundred planes to be operated as a unit, and two separate brigades, one along the Atlantic Coast and one along the Pacific Coast, of six hundred airplanes each. The force should consist of two-thirds pursuit and one-third bombardment (included in the bombardment aviation should be some attack aviation, if this continues to be an element in our organization).

There should be a local defense unit of one hundred planes and accessories to defend the City of New York, to be used as a model on which defense against aircraft organizations for other places could be based when required.

There should be a local defense unit of one hundred airplanes at Panama. The airways should be organized throughout the United States, Porto Rico, the West Indies Islands, Cuba, Mexico, and Central

America to Panama, so that air force units could be dispatched to the most important point for the defense of that area.

The Hawaiian Islands, due to their remoteness from the continent, should be equipped with an air force of three hundred airplanes, two-thirds of which should be pursuit and one-third, bombardment. In addition to the air force units, there should be a unit for the local defense of the Island of Oahu of one hundred pursuit planes and accessories for defense against aircraft.

Due to the strategic position of the Philippine Islands, there should be no air force or local defense units maintained because the locality could not be defended in case of war. The aeronautical organization there should consist merely of two general service squadrons of twenty-five two-seater airplanes each. These should be organized for use against local uprisings, for reconnaissance, and in developing the airways of the Islands.

Alaska should be provided with an air force of three hundred planes, two hundred pursuit and one hundred bombardment. These should have their headquarters in time of peace in the vicinity of Fort Gibbon, opposite the mouth of the Tanana River on the Yukon. The airways should be organized, with the consent of Canada, from the United States to Alaska as far as Nome and Cape Prince of Wales, and also down the Alaskan Peninsula and the Aleutian Islands to the Island of Attu.

Of the force in the continental United States, there should be about fifteen per cent of officers and enlisted men maintained permanently with the organizations and the rest in Reserve. The general overhead, engineering and supply services should be permanent. In Panama, the units should remain at half strength; in the Hawaiian Islands, they should be at full strength; in the Philippines, at full strength; and in Alaska, at half strength. The number and strength of both the local defense and auxiliary units for observation purposes should be varied from time to time in accordance with the needs of the organizations which they are designed to serve.

The greatest necessity exists for the creation of an air personnel as distinguished from an Army or Navy personnel. The Air Service suffers annually nearly half of the total deaths in the Army: in 1921, 42%; 1922, 43%; 1923, 47%; 1924, 41%. In war, the proportion of fatalities among the commissioned personnel is even greater. This requires an entirely different system of replacement of personnel from that followed in the Army or Navy, a different Reserve system, and an entirely different system of entrance into the service, promotion, and retirement.

The Air Service navigating personnel should be drawn from the younger elements in our population that are well educated, of an athletic disposition, and fit physically for the duties. Upon entrance into the service, they should be guaranteed a certain promotion, based on years of service. Those who distinguish themselves should be placed in command of organizations and be given the temporary rank which the duties and responsibilities of the position require.

Those that are retired for reasons not due to their own fault, should be given retired pay commensurate with their years of service. A system of this kind would guarantee a career to the individual and would also provide a suitable personnel for the various units in the Air Service.

Unity of command is essential to air forces. These can not be operated efficiently in time of war if scattered and assigned to ground or water organizations.

The system of education of air officers primarily as officers of the ground army is wrong. It tends to promote timidity of operation, lack of foresight as to air needs, and lack of ability to lead air troops. Air personnel should be educated and brought up essentially at air institutions of learning. Liaison with other branches of the service should be reciprocal and the personnel of the ground army should be instructed in the duties of air forces by air officers instead of by ground officers. This is a very serious defect in our system which will continue to grow worse because our officers are now educated primarily on the ground and secondarily in the air.

The present system of budget control for the Air Service is destructive of the development of air power. Air forces have no relation to ground forces any more than a navy has. Air forces must be designed primarily to attain victory in the air against a hostile air force and then to destroy enemy establishments, either on the land or on the water. Air power should have an entirely separate budget from the Army and Navy.

Under the conditions that exist today where military

aeronautics are a part of the Army and Navy, aviation cannot obtain the consideration necessary to meet its requirements of building up its offensive aviation for wresting control of the air from an enemy, because the basic principle is followed that an army must be built up on infantry and a navy on battleships.

Unity of tactical instruction is necessary. At the present time, none exists between the air forces of the army or of the navy.

In practically all of the civilized countries of the world there is unity of command in the air forces. They are handled by general officers of the air forces. The separate and independent fighting air units are directly under the tactical and strategical command of the Commander in Chief of the respective countries. In sharp distinction to this, there is no single command in the U. S. aviation, nor is there actually any air force in the United States.

The time has passed when any one service can be thrown off to work out its own salvation without respect to the others, as has been the case very largely in the past with the armies and navies. Air, land and water must be hitched together under one general command and direction to provide for an efficient defense.

As a result of many years of service and an intimate knowledge of the aeronautical organization of each of the great powers I am convinced that our inefficient national military aeronautics, our undeveloped civil and commercial aeronautics, and our curtailed and interrupted experimentation in aeronautics are a direct result of the lack of

1. A department of aeronautics to handle the whole air question, co-equal with the Army and Navy.
2. A definite aeronautical policy.
3. An organization, both military and civil, to fit the aeronautical policy.
4. A method of providing suitable personnel for all air undertakings.
5. A single system of procurement and supply for all air undertakings.
6. A system of instruction and inspection for all air elements.

Until these fundamental principles for the creation of air power are put into effect, the air power of the United States will continue to flounder in the slough of aeronautical despond.

# APPENDIX

## THE UNITED STATES GOVERNMENTAL AGENCIES CONCERNED WITH AERONAUTICS AND WHAT THEY ARE DOING

This statement of the duties of the executive branches of the Government charged with aviation activities was published by the National Advisory Committee for Aeronautics in 1924. It gives quite a good insight into the *disorganization* of our national aeronautics. The reader will gain a good idea of what each branch of the Government is doing with aviation, of its complication, lack of coordination and the impossibility of handling it efficiently in case of war or of developing it in time of peace.

The Governmental agencies mentioned herein are not the only ones that use aeronautics in one way or the other; practically every branch of the Government has something to do with it.

There are four governmental agencies directly concerned with the use or development of aviation, namely:

The Army Air Service.
The Naval Bureau of Aeronautics.
The Air Mail Service.
The National Advisory Committee for Aeronautics.

### THE ARMY AIR SERVICE

The Army Air Service was established in its present form by the Army reorganization act, approved June 4, 1920, and at present functions under the control of the Secretary of War as a coordinate branch of the Army.

The functions of the Army Air Service have been classified first, as an arm of the mobile army; second, as an arm to be used against enemy aircraft in defense of all shore establishments; and third, as an arm to be used in cooperation with other arms, or alone, against enemy vessels engaged in attacks on the coast.

The Chief of the Army Air Service has the rank of major general. The organization is divided into six main divisions, namely, personnel, information, training and war plans, industrial war plans, supply, and engineering.

The flying personnel of the Army Air Service at the present time is obtained by the assignment of graduates from West Point, by the transfer of junior officers from other arms, and by appointment after examination of applicants from civil life.

The Air Service has 845 officers with rating as airplane pilots, airplane observers, airship pilots, airship observers, or balloon observers. In addition about 51 enlisted men have the rating of airplane pilot, junior airplane pilot, or airship pilot.

The following special stations are maintained by the Army Air Service and have the functions specified.

Brooks Field, San Antonio, Tex. (the primary flying school).
Chanute Field, Rantoul, Ill. (the technical school).
Kelly Field, San Antonio, Tex. (the advanced flying school).
Langley Field, Hampton, Va. (the tactical school).
McCook Field, Dayton, Ohio (the engineering school).
Scott Field, Belleville, Ill. (the balloon and airship school).

In addition there are a number of other fields occupied by tactical units which are under the immediate command of the corps area commanders in the United States and under the department commanders in Hawaii, the Philippine Islands, and the Canal Zone. These units are so located that at all times the ground forces of the Army may have the aerial observation cooperation so essential in their peacetime training and when engaged in hostile operations. As a small nucleus for the development of an adequate air force capable of fulfilling that part of the mission of the Army Air Service as quoted above "as an arm to be used against enemy aircraft in defense of all shore establishments; and as an arm to be used in cooperation with other arms, or alone, against enemy vessels engaged in attacks on the coast," the Army maintains, exclusive of overseas garrisons, one bombardment, one attack, and one pursuit group at Langley, Kelly, and Selfridge Fields, respectively. These tactical air force units are engaged in the development and perfection of offensive and defensive aerial tactics, air strategy, and the concentrated application of tactical air power. With the limited forces available air defense maneuvers have only been possible of accomplishment on a small scale. The Army Air Service has repeatedly proved the value of aviation as a peace-time agency by successfully employing it in the following activities of a nonmilitary nature: Forest fire patrol; aerial survey and mapping; in conjunction with the Department of Agriculture in combating insect pests, such as the boll weevil and gypsy moth, and in other ways.

The engineering division of the Air Service at McCook Field, Dayton, Ohio, carries on engineering experiments covering the development of airplanes, engines, armament, materials, instruments, parachutes, studies of design possibilities, night flying equipment, etc. Briefly, the engineering division conducts engineering experiments which are of direct value to the Army Air Service and in the general development of aviation. Airplanes are not manufactured by any Air Service station.

Equipment to demonstrate the practicability of night flying was developed at McCook Field and tested by 7,500 miles of night flying between Dayton and Columbus, Ohio. The results of these tests formed the basis upon which the Air Mail Service inaugurated its present successful night-flying service.

The Air Service has developed aerial photography so that it is being used by many departments of the Government. In addition to its tactical application by all branches of the Army, aerial photographs of large areas of the

United States have been taken for the Geological Survey and Corps of Engineers. All such information gathered by the Air Service is made available to commercial organizations.

The Air Service has started and operated airways between some of the principal cities of the United States and has prepared data on air routes and landing fields, so that when commercial flying reaches a stage where this information is needed it can be furnished.

The Air Service also trains a number of flying cadets each year who are given reserve commissions and are thereafter available as military pilots or are well qualified to be pilots of civil aircraft.

A number of interesting and important flights, all having a military significance, have been made by the Air Service. Among these may be mentioned the Alaskan flight, various transcontinental flights, the Porto Rican flight, the dawn-to-dusk flight, and the round-the-world flight. These flights have served their purpose and have brought home to the American people some of the possibilities of aviation.

## THE NAVAL BUREAU OF AERONAUTICS

The Naval Bureau of Aeronautics was established by act of Congress approved July 12, 1921. Its functions differ from those of the Army Air Service, due to a fundamental difference in organization, in that the Army Air Service is a combatant arm of the Army with its own production and supply services, etc., whereas the Navy has no separate combatant arms, naval aviation being an integral part of the fleet. The Naval Bureau of Aeronautics is charged with all matters relating to the design, procurement, development, and maintenance of naval and Marine Corps aircraft, and the carrying into effect of the Navy Department's policies regarding naval aviation. The Naval Bureau of Aeronautics furnishes all information concerning naval aviation required by the Chief of Naval Operations, who, under the direction of the Secretary of the Navy, is charged with the operations of the fleet (including aircraft) and with the preparation and readiness of plans for its use in war.

Flying personnel and aircraft units forming the naval aviation organization afloat are under the immediate command of the commanders of the vessels to which they are attached when such vessels are acting singly, and they are under the immediate command of the senior officer present when such vessels are acting collectively. The tactical command of aircraft in flight is exercised by the senior flying officer. The mission of such aircraft is to serve the needs of the ships to which they are attached.

Provision is also made for fleet aircraft squadrons operating from airplane carriers or tenders under the immediate command of the fleet aircraft squadron commander, who is responsible to the commander in chief of the fleet. Fleet aircraft squadrons operate as parts of the fleet in the same manner as other coordinate arms of the fleet—surface craft and sub-surface craft.

Another part of the naval organization provides for aircraft assigned to the naval districts as a part of the naval coast defense forces operating under the various district commanders ashore.

The Chief of the Naval Bureau of Aeronautics has the rank of rear admiral. The organization is divided into four main divisions, namely, plans, administration, material, and flight.

Naval flying officers are selected at the present time solely from line officers of the Navy and of the Marine Corps. The present requirements are that they must be graduates of Annapolis and have had three years of sea duty before being selected for aviation duty. There are, however, a number of flying officers who entered the naval service during the war direct from civil life who have since qualified by examination as regular line officers of the Navy.

There are 630 officers attached to naval aviation, including 157 ground officers. Of the total of 630 officers, 422 are attached to naval aviation ashore and 208 are attached to naval aviation afloat.

The following special stations have the functions stated:

Naval Air station, Anacostia, D. C. (experimental and test work).
Naval air station, Lakehurst, N. J. (rigid airship operation and maintenance.)
Naval air detail, Newport, R. I. (experimental torpedo work).
Naval air detail, Dahlgren, Va. (experimental ordnance work).
Naval aircraft factory, navy yard, Philadelphia, Pa. (repair and maintenance; experimental and test work; storehouse and supply depot).
Naval air station, Pensacola, Fla. (training of flying personnel).
Aviation mechanical school, Great Lakes, Ill. (training of aviation mechanics).

In addition to the above, the Navy has four active and three inactive coastal air stations.

The principal activities of the Bureau of Aeronautics at the present time may be summarized as follows: Development of service types of airplanes, aircraft engines, and accessories, and their procurement, supply, and maintenance; providing naval aircraft units with the latest approved types of airplanes and equipment; development of rigid airships by assignment of the Joint Army and Navy Board, and of other lighter-than-air activities; installing aircraft on vessels of the fleet; development of aircraft carriers; development of launching and arresting devices for airplane carriers and shipboard use; training of regular and reserve aviation personnel; aerial photography and mapping; aerological reports; maintenance of naval air stations ashore; and cooperation with the Army Air Service, the National Advisory Committee for Aeronautics, and, as far as possible, with civilian aeronautic organizations, for the furtherance of aviation development.

## THE AIR MAIL SERVICE

The Air Mail Service was inaugurated May 15, 1918, the first route being between Washington and New York.

It has been supported by annual congressional appropriations without having been definitely established by law.

It is a transportation service directly operated by the Post Office Department under the immediate control of the Second Assistant Postmaster General.

The personnel of the Air Mail Service totals 580, including 42 regular airplane pilots and 5 reserve pilots. The pilots, as well as the other personnel, are secured direct from civil life without examination.

The flying equipment of the Air Mail Service comprises a total of 94 airplanes, of which there are 82 DH-4's used for carrying the mails, 4 inspection airplanes in good condition, and 8 others that are not serviceable.

The Air Mail stations in operation number 15, extending across the country on the route from New York to San Francisco. There is also an Air Mail general repair depot located at Chicago, employing 115 men. At this station airplanes are overhauled and rebuilt, and spare parts are stocked for all flying equipment and ground equipment, especially that needed for night flying.

In a special report of the National Advisory Committee for Aeronautics submitted to President Harding on December 20, 1922, at his request, it was stated that:

"The fundamental purpose of the Air Mail Service is to demonstrate the safety, reliability, and practicability of air transportation of the mails, and incidentally of air transportation in general. In particular, it should—

"(a) Develop a reliable 36-hour service between New York and San Francisco, and make that service self-supporting by creating the necessary demand for it and charging a rate between ordinary postage rates and night-letter telegraph rates.

"(b) Keep strict records of the cost of the service and strive in every way to reduce such costs to a minimum, thereby demonstrating the value of air transportation from an economic point of view, and in particular making it possible for private enterprise eventually to contract for the carrying of mails by airplane at a rate which not only would not exceed the income from such a service but would permit the Post Office Department to provide other postal airways to meet the demands of the people for the more rapid transportation of mail. In the present undeveloped state of the art, it would be wholly impracticable to operate an air mail service by contract."

The Air Mail Service at the present time is conducting an experimental demonstration of the practicability of night flying in the transportation of mail between New York and San Francisco by air. The ground equipment for night flying extends from Bryan, Ill., near Chicago, to Rock Springs, Wyo., near Cheyenne, and mail is being transported regularly on an approximate average of 41 hours for westbound mail and 36 hours for eastbound. The use of the service is gradually increasing, and the developments to date indicate that in a reasonable time the service will be fully self-supporting.

The development of the Air Mail Service has been a credit to American aviation. It is a practical means for aiding the development of commercial aviation as well as a means for expediting the transportation of mail. Mail is bound to be carried eventually by the fastest means available, and it is safe to say that in this age of progress the American people will demand and will support a more or less general use of aircraft in the future for carrying the mails.

## THE NATIONAL ADVISORY COMMITTEE FOR AERONAUTICS

The National Advisory Committee for Aeronautics was created by act of Congress in 1915 as an independent establishment. It is composed of 12 members appointed by the President, all of whom serve without compensation. Its membership is drawn from official and private life as follows:

From the Government Service:

Two from the War Department (the Chief of Air Service and the chief of the engineering division of the Air Service.)

Two from the Navy Department (the Chief of the Bureau of Aeronautics and the chief of the material division).

One from the Weather Bureau (the chief).

One from the Bureau of Standards (the director).

One from the Smithsonian Institution (the Secretary).

From private life:

Five who are acquainted with the needs of aeronautical science or skilled in aeronautical engineering or its allied sciences.

The organic act creating the National Advisory Committee for Aeronautics provides: "That it shall be the duty of the Advisory Committee for Aeronautics to supervise and direct the scientific study of the problems of flight, with a view to their practical solution, and to determine the problems which should be experimentally attacked, and to discuss their solution and their application to practical questions. In the event of a laboratory or laboratories, either in whole or in part, being placed under the direction of the committee, the committee may direct and conduct research and experiment in aeronautics in such laboratory or laboratories."

The committee's laboratories for the direct conduct of fundamental research in aeronautics are located at Langley Field, Va., where the facilities of a large Army flying field are added to those of a well-equipped research laboratory.

The committee operates under rules and regulations approved by the President. It elects annually its chairman and its secretary from among its members and also an executive committee, which in turn elects its chairman and its secretary. The executive committee has immediate and entire charge of the activities of the committee during the interim between the stated meetings of the entire committee.

The executive committee has established three standing technical sub-committees, namely, the committee on aerodynamics, the committee on power plants for aircraft, and the committee on materials for aircraft. The organization of these subcommittees is patterned after that of the entire committee, each having specially appointed representatives from the Army and Navy Air Services, the Langley Memorial Aeronautical Laboratory, the Bureau of Standards, and private life, all of whom serve without compensation.

It is mainly through the instrumentality of the technical subcommittees that coordination of aeronautic research and experiment and cooperation among the agencies interested are made effective. The subcommittees originate the programs for aeronautical research in their respective fields, and after such programs are approved by the executive committee the subcommittees receive progress reports periodically and keep in touch with the active workers in their respective fields and with the progress of the investigations coming under their cognizance. Through these subcommittees, and largely by virtue of the opportunity afforded by regular and official contact and personal acquaintance that result from regular attendance at meetings, complete cooperation on the part of all responsible governmental officers concerned with the investigation of technical problems in aeronautics is assured.

With the subcommittees functioning efficiently, and with their activities coordinated by the director of aeronautical research, the executive committee is enabled to devote a portion of every meeting to the informal discussion of general problems regarding the development of military and civil aviation, and these informal discussions are often of greater advantage in promoting understanding and cooperative effort than are formal or official communications.

The committee's activities in the field of aeronautical development may be stated under four headings as follows:

(a) The coordination of research and experimental work in aeronautics by the preparation of research programs and the allocation of particular problems to the various laboratories.

(b) The conduct of scientific research on the more fundamental problems of flight, under the immediate direction of the committee in its own laboratory known as the Langley Memorial Aeronautical Laboratory, at Langley Field, Va.

(c) The collection, analysis, and classification of scientific and technical data in aeronautics, including the results of research and experimental work conducted in all parts of the world.

(d) The diffusion of technical knowledge on the subject of aeronautics to the military, naval, and postal air services, aircraft manufacturers, universities engaged in the teaching of aeronautics, and the public generally.

In addition, the committee holds itself at the service of the President, the Congress, and the executive departments of the Government for the consideration of special problems in aeronautics which may be referred to it.

The success of the National Advisory Committee for Aeronautics in performing its important functions depends fundamentally upon the following facts:

1. Its members and the members of its standing subcommittees serve without compensation, the Government thus receiving the services of men who would not otherwise be available.

2. The reason why these men are willing to serve is because the committee is an independent Government establishment, reporting directly to the President, receiving its own appropriation from Congress.

3. By virtue of such status, the committee is able to initiate and to conduct any investigation which, after full discussion by the subcommittee concerned, is considered fundamental or desirable.

All these advantages would be lost were the committee to be made part of any Government department.

## THE EXISTING SCHEME OF COOPERATION

### Organization

The governmental agencies for effecting cooperation and preventing duplication are:

The Joint Army and Navy Board.
The Aeronautical Board.
The National Advisory Committee for Aeronautics.

### The Joint Army and Navy Board

The Joint Army and Navy Board was created in 1919 by joint order of the Secretary of War and the Secretary of the Navy "to secure complete cooperation and coordination in all matters and policies involving joint action of the Army and Navy relative to the national defense." The board is composed of three high-ranking Army officers and three high-ranking Navy officers, including the Chief of Staff of the Army and the Chief of Naval Operations. The order creating the board further provides that "It shall also have the duty of originating consideration of such subjects when in its judgment necessary; and is responsible for recommending to the Secretary of War and the Secretary of the Navy jointly whatever it considers essential to establish the sufficiency and efficiency of cooperation and coordination of effort between the Army and the Navy."

In order to provide an agency for detailed investigation, study, and development of policies, projects, and plans relative to the national defense and involving joint action of the Army and Navy, the Secretary of War and the Secretary of the Navy have further agreed:

Upon the organization of a joint Army and Navy planning committee, consisting of—

For the Army: Three or more members of the War Plans Division, General Staff, to be designated by the Chief of Staff.

For the Navy: Three or more members of the Plans Division of Naval Operations, to be designated by the Chief of Naval Operations—

the order establishing the Joint Army and Navy Board provides that "the joint Army and Navy planning committee will investigate, study, and report upon questions relative to the national defense and involving joint action of the Army and Navy referred to it by the Joint Army and Navy Board. It shall also have the duty of originating consideration of such subjects when in its judgment necessary. The members of this committee are authorized to consult and confer freely on all matters of defense and military policy in which the Army and the Navy are jointly concerned and will consider this joint work as their most important duty."

### The Aeronautical Board

The Aeronautical Board was created in 1920 by joint action of the Secretary of War and the Secretary of the Navy, and is composed of three officers from the Army Air Service, and three officers from the Navy Bureau of Aeronautics. The following paragraphs are quoted from the order establishing the Aeronautical Board:

"To prevent duplication and to secure coordination, plans of new projects for the construction of aircraft, for experimental stations, for coastal air stations, and for stations to be used jointly by the Army and Navy, or for extensive additions thereto, shall be submitted to the Aeronautical Board for recommendation.

"The development of new types of aircraft, or of weapons to be used from aircraft, so far as practicable, shall be assigned to and carried on by one Air Service. This restriction shall not prevent the employment by either Air Service of any types of aircraft or weapons which, after development, are considered to be necessary for the accomplishment of its functions. Questions relating to the development of new types of aircraft or weapons to be used from aircraft shall be referred to the Aeronautical Board for recommendation as to which Air Service shall be charged with the development.

"All information pertaining to experiments in connection with aviation shall be exchanged promptly between the Army and Navy Air Services.

"Whenever possible, training and other facilities of either Air Service shall be made available for, or to be used by, the other service.

"In the interests of economy, heavier-than-air craft shall be provided in preference to nonrigid, semirigid, or rigid dirigibles whenever the former can satisfactorily perform the service required.

"All estimates of appropriations for the Army and Navy aviation programs shall be presented to the Aeronautical Board for review and recommendation before submission to Congress."

### The National Advisory Committee for Aeronautics

The National Advisory Committee for Aeronautics functions under the President as an independent Government establishment created by act of Congress in 1915, for the supervision and direction of the scientific study of the problems of flight, with a view to their practical solution. It conducts scientific research on the more fundamental problems of flight, and in the exercise of its functions as a coordinating agency it allocates to other agencies, governmental and private, investigations in aeronautics for which they may be peculiarly fitted, thus marshaling the talent of the American people for the advancement of the science of aeronautics.

The regular and official contact, through membership on the main committee and on its various subcommittees, of representatives of the Army and Navy Air Services with each other and with specially appointed representatives of the Bureau of Standards, educational institutions, and independent enterprises that are aiding in the solution of technical problems in aeronautics, serves to prevent unnecessary duplication of effort, facilitate the interchange of ideas between different groups of workers, and develops a mutual understanding that makes for the greatest possible advancement in the science of aeronautics.

The committee's unique facilities for aeronautical research are also utilized in the making of special investigations on request of either the Army Air Service, the Naval Bureau of Aeronautics, or the Air Mail Service, and the same facilities are available for the conduct of special investigations for private firms or individuals, provided that they defray the actual cost thereof.

The committee invites special attention to the fact that the Bureau of Standards has been of great assistance in the study of technical problems in aeronautics for which it had adequate facilities. The same may be said of the Weather Bureau in the study of meteorological problems and their relation to aviation; of the Forest Products Laboratory in the study of wood problems; and of the Bureau of Mines in the development of helium. It has been the policy of the committee to assign to the Bureau of Standards investigations for which it was peculiarly well equipped, and the committee has encouraged the use of that bureau's facilities by the Army and Navy Air Services. The facilities of educational institutions, notably the Massachusetts Institute of Technology and Stanford University, have also been utilized by the Government to advantage in the investigation of aeronautic problems.

## POLICY OF THE ARMY AND NAVY RELATING TO AIRCRAFT

The following policy of the Army and Navy relating to aircraft has been approved by the Joint Army and Navy Board and by the War and Navy Departments, and has been published to the services for their information and guidance.

"Aircraft to be used in the operations of war shall be designed:

"(a) Army aircraft.
"(b) Navy aircraft.
"(c) Marine aircraft.

"Army aircraft are those provided by the War Department and manned by Army personnel.

"Navy aircraft are those provided by the Navy Department and manned by Navy personnel.

"Marine aircraft are those provided by the Navy Department and manned by Marine Corps personnel.

"The marine air service is a branch of the naval air service.

"The functions of the Army, Navy, and marine aircraft are as follows:

"*Army aircraft.*—Operations from bases on shore—

"(a) As an arm of the mobile army.
"(b) Against enemy aircraft in defense of all shore establishments.
"(c) Alone or in cooperation with other arms of the Army or with the navy, against enemy vessels engaged in attacks on the coast such as—
"I. Bombardment of the coast.
"II. Operations preparatory to or of landing troops.
"III. Operations such as mine laying or attacks on shipping in the vicinity of defended ports.

"*Navy aircraft.*—Operations from mobile floating bases or from naval air stations on shore—
"(a) As an arm of the fleet.
"(b) For overseas scouting.
"(c) Against enemy establishments on shore when such operations are conducted in cooperation with other types of naval forces, or alone when their mission is primarily naval.
"(d) To protect coastal sea communications by—
"I. Reconnaissance and patrol of coastal sea areas.
"II. Convoy operations.
"III. Attacks on enemy submarines, aircraft, or surface vessels engaged in trade prevention, or in passage through the sea area.
"(e) Alone or in cooperation with other arms of the Navy, or with the Army, against enemy vessels engaged in attacks on the coast.

"*Marine aircraft.*—The functions normally assigned to Army aircraft shall be performed by the marine aircraft when the operations are in connection with an advance base in which operations of the Army are not represented. When Army and marine aircraft are cooperating on shore, the control of their operations shall be governed by the one hundred and twentieth article of war, United States Army.

"The functions of aircraft assigned under Army (c) and Navy (e) are a duplication of functions. In such operations cooperation is vital to success. Such cooperation shall be governed by the following provisions:

"(a) The naval district forces, vessels and aircraft, will never be strong enough to prevent an attack on the coast by major units of the enemy fleet. When, therefore, an enemy force of a strength greatly superior to that of the naval force available for use against it approaches the coast the commander of the naval force shall inform the commander of the Army department of the situation; shall assume that the Army has a paramount interest in the operation, and shall coordinate the operations of the naval forces with those of the military forces.

"(b) If, however, the conditions are such that the enemy is, or can be, engaged by a naval force approximating in strength to that of the enemy, the commander of the Army department shall be so informed and he shall assume the Navy has a paramount interest in the operation and shall coordinate the operations of the military forces with those of the naval forces.

"The functions of aircraft above assigned shall govern in the production of aviation equipment, training of aviation personnel, and establishment of air stations for the War and Navy Departments. Such assignment shall not prevent the employment of Army and marine aircraft in naval functions upon the request of the senior naval officer present or vice versa, the employment of naval aircraft in Army functions upon request of the senior Army or marine officer present on shore; nor shall it prevent the employment of Army, Navy, or marine aircraft when no other Air Service is cooperating in the operation, in any manner which shall be most effective in accomplishing the mission of the force.

"All questions regarding the policy of the War and Navy Departments with regard to the tactical and strategical functions of aircraft, and to the location of air stations, shall be addressed to the joint board for consideration and recommendation to the Secretary of War and the Secretary of the Navy."

## THE INCREASING IMPORTANCE OF AIRCRAFT IN WARFARE

The Limitations of Armaments Conference, held in Washington in 1921-22 on invitation of President Harding, examined into the possibility of limiting aviation development for war purposes and limiting the use of aircraft in warfare. A special committee of aviation experts representing the United States, Great Britain, France, Italy, and Japan was appointed. That committee submitted a report which reviewed the situation at length. The "final conclusions" of the report follow:

"The committee is of the opinion that it is not practicable to impose any limitations upon the numbers or characteristics of aircraft, either commercial or military, except in the single case of lighter-than-air craft. The committee is of the opinion that the use of aircraft in warfare should be governed by the rules of warfare as adapted to aircraft by a further conference which should be held at a later date."

The fact that the Limitation of Armaments Conference placed no restriction on the development and application of aircraft for war purposes assures the greater relative importance of aircraft in future warfare. It is a maxim of military science that an army and a nation must be adaptable to changes in time of war. The best laid plans, whether for offensive or defensive warfare, are usually upset either by the success of the enemy or by changes and developments in the art of warfare. No one can foretell at this time what the use of aircraft will be in future wars, nor even in the next war. It is safe to say that there will be individual and group fighting in the air; there will be aircraft attacking troops on the ground, both with bombs dropped from great heights and with machine guns mounted on low-flying aircraft protected by armor from ordinary rifle bullets; there will be bombing of large cities, military and manufacturing centers, and routes of communication and transportation. And it has been proposed that aircraft be used to drop poisonous gases not only on to the enemy's troops but also behind the lines and in the centers of population, to the same extent that long-distance bombing will be carried on. The bombs carried may not be limited to explosives and poisonous gases but may possibly be loaded with germs to spread disease and pestilence. Without limitations on the uses of aircraft in warfare a nation fighting with its back to the wall can not be expected to omit to use desperate means to stave off defeat. The uses of aircraft in warfare would then be limited only by the inability of human ingenuity to conceive further uses for this new agency of destruction.

A conference was held at the Hague in 1923, attended by delegates of the United States, which drafted rules and regulations covering the use of aircraft in war. There was evident a tendency to minimize as much as possible aircraft attacks upon centers of population with the resulting consequences to noncombatants, and to restrict such attacks to what are military objectives. In spite of the rules thus formulated, and even if they should be universally adopted, it is still inevitable that aircraft attacks would greatly terrify and undoubtedly seriously injure and damage many who have heretofore been classed as noncombatants.

It is believed quite probable that if the nations of the world do maintain adequate air forces, this may tend to the adjustment of international disagreements by conference, as the delegates to such conferences will have the strong backing of their national air forces capable of such destructive effect as that indicated above. When wars were fought within a limited territory by ground troops, the national patriotism of noncombatants strongly supported their armed forces, but in future wars when air power becomes a most vital factor in national defense, theaters of operation will no longer be limited to restricted territories, and noncombatants will probably and unavoidably be subjected to far greater personal danger and injury than in the past. It is not inconceivable that such pressure will be brought to bear upon the Governments concerned by their noncombatants, following a series of aircraft raids, that an early cessation of hostilities will be more earnestly desired by the people on both sides and will be forced by popular demand upon the nation least efficient in air power.

Aviation has made itself indispensable to military and naval operations. Under our present organization, where the function of national defense is vested principally in the War and Navy Departments, we must look to those departments to develop the possibilities of aviation in warfare, whether to be used in conjunction with military and naval operations, or to be used independently for attacking distant points behind the enemy's lines, or elsewhere. The problem of the air defense of this country is worthy of most careful study.

## RELATION OF AERONAUTIC RESEARCH TO NATIONAL DEFENSE

So long as the development of aviation continues from year to year, the military and naval policies and programs for our national security and defense are necessarily subject to change, as they are largely dependent upon the probable use of aviation in future wars. So long as other nations are seriously engaged in the development of aviation, America must at least keep abreast of the progress of aviation abroad and never permit itself again to fall behind as it did before the World War. Substantial progress in aviation, whether in America or elsewhere, is in the last analysis dependent upon aeronautical research. It is necessary that accurate information, which is the result of scientific research on the fundamental problems of flight, should be made continuously available to the Army and Navy; and those agencies desire from the National Advisory Committee for Aeronautics the fundamental aerodynamic information on which the design of new types of military and naval aircraft is based. It is the function of the Army and Navy then to check this information and apply it in an engineering manner to the design of aircraft.

While national defense is the greatest use to which aircraft is applied in America to-day, the committee believes that the time will come when its military uses will be second in importance to its civil value in promoting our national welfare and increasing our national prosperity. But to-day, while the uses of aircraft are primarily military and the Air Services of the Army and Navy are not as large as those of other world powers, America is gradually forging ahead of other nations in the acquisition of knowledge of the scientific principles underlying the design and construction of aircraft. To this important but limited extent, we are providing well against unpreparedness in the air.

## THE AIRCRAFT INDUSTRY AND ITS RELATION TO NATIONAL DEFENSE

The present American industry is but a shadow of that which existed at the time of the armistice. With the great stimulus in aircraft development and performances during the war, the aircraft manufacturers were hopeful that civil aviation would rapidly come into being with a resulting great demand for their product. Civil aviation has not developed as it was hoped it would, and this makes the present situation more difficult.

These aircraft manufacturers have had to rely for orders upon Government agencies, and the limited amount of governmental purchases has forced a number of manufacturers to go out of the aircraft business. It is a matter of grave Government concern lest the productive capacity of the industry may become so far diminished that there may not remain a satisfactory nucleus. By a "satisfactory nucleus" is meant a number of aircraft manufacturers, distributed over the country, operating on a sound financial basis, and capable of rapid expansion to meet the Government's needs in an emergency.

After the very costly lessons of the war, it would be

folly to say that the Government is not concerned with the state of aircraft industry. It is concerned that there should be in existence, and in a healthy condition, at least an adequate nucleus of an industry. An aircraft industry is absolutely essential to national defense. One lesson of the war that will not be forgotten is that it takes a great deal of money to develop hastily an aircraft industry from almost nothing. The American people can ill afford to pay such a price a second time. To maintain a nucleus of an industry it has been proposed either that the Government substantially increase the volume of its orders for aircraft, or devise a policy for the apportionment of orders at fair negotiated prices without regard to competition.

Neither of these propositions, however, in the judgment of the National Advisory Committee for Aeronautics, goes to the root of the trouble. To substantially increase orders will require substantially increased appropriations. To increase appropriations for the Army and Navy Air Service because they need more aircraft is one thing, but to increase appropriations primarily to maintain an aircraft industry is something else. Furthermore, the maintenance of an industry in a healthy condition does not involve the maintenance of any manufacturer who has failed to liquidate or reduce his plant and overhead expenses to an appropriate peace-time basis.

In the judgment of the committee, the existing bad situation in the industry should be substantially remedied. In an effort to help the situation, the committee suggests the following steps on the part of the industry and of the Government:

*Steps to be taken by the industry:*

First. Every manufacturer intending to remain in the aircraft business and who has not readjusted his war-time plant and overhead expenses to a peace-time basis should do so without further delay.

Second. The firms comprising the aircraft industry should specialize in the production of various types of aircraft with a view to the more continuous development of types by the same plants and the gradual recognition of proprietary rights in new designs.

*Steps to be taken by the Government:*

First. The Army, Navy, and Postal Air Services should agree upon a balanced program setting forth from time to time the probable requirements of the Government for each type of aircraft for at least one year in advance, and should announce the same to the industry for its information and guidance.

Second. Orders for the different types should be placed with the different manufacturers at such intervals as to insure continuity of production and the gradual development of special facilities and skill by each manufacturer in the production of a given type of aircraft.

The committee does not attempt to say that the method proposed is the ideal solution, but it submits that if followed it would produce the following beneficial results:

(a) It would insure the continuous development of types by the same firms which is the most rational method of improving the quality and performance of aircraft to meet special needs.
(b) It would reduce the cost of aircraft.
(c) It would provide all manufacturers with an adequate market to enable them to continue in the airplane business without the periodical menace of dissolution or bankruptcy heretofore caused by long gaps, between orders.

## COMMERCIAL AVIATION AND ITS RELATION TO THE GOVERNMENT

The stimulus of war forced the development of aviation for military purposes, and while the progress thus made was beneficial to all aviation, nevertheless there has been little application of aviation to commercial purposes. In England, France, Italy, Germany, Holland, Belgium, Poland, and other European countries there are air lines for the transportation of passengers and goods on regular schedules across international boundaries and intervening seas. It is quite a customary thing for tourists and business men to travel by air, for example, between London and Paris. There is a great rivalry for business between French and English companies, all of which are subsidized by their Governments.

There is at the present time in the United States no large regular air transportation business, although enterprising firms from time to time have undertaken to establish more or less regular routes between points deemed peculiarly attractive for the development of an air transportation business. The Air Mail Service operated by the Post Office Department has given the best and most practical demonstration of the reliability and adaptability of aircraft to the useful purposes of commerce. The present experiment by the Air Mail Service to determine the practicability of night flying is the most important development in aviation to-day and should prove to be of substantial assistance in the development of commercial aviation in America.

The reason for the greater development of commercial aviation in European countries to date lies in the fact that they realize more keenly than we in America do the vital necessity of aviation to national defense. They are either adjoining neighbors or within a few hours of each other by air, and unless military aviation in those countries is to bear the entire cost of the maintenance of aircraft industries and of aviation development generally, those countries must in sheer self-defense encourage commercial aviation. This they have done in every practicable way, principally by subsidizing common carriers by air, especially those engaged in international aerial transport.

In the United States direct subsidy appears to be out of the question because of our adherence to a traditional policy. In our country aviation must make its own way. Civil aviation has not progressed very far because it has not yet reached that stage of development that justifies its use generally from an economic point of view, unless an inordinate value is to be placed upon speed. Speed and maneuverability may be prime factors in military aircraft, especially in time of war, but for commercial purposes aircraft must be made safer, more controllable at low speeds incident to taking off and landing, and less expensive in initial cost as well as in maintenance and operation.

Commercial aviation will have to be regulated, just as are other means of transportation. The initial legislation in this respect should be very carefully prepared, so that, while affording that degree of regulation considered necessary in the public interest and that degree of practical assistance that would be helpful, it will nevertheless leave the new art of aviation ample freedom to develop normally without unnecessary or unwise restrictions and without attempting to set up by legislature an artificial basis for the maintenance of the activity to be regulated.

## SUMMARY

Aviation has been proved indispensable to both the Army and the Navy. Neither one can operate effectively without an adequate air service. What was considered adequate in the World War will not do in the future.

The progress in scientific research and in the technical development of airplanes and airships has been continuous and gratifying; but before commercial aviation can become self-supporting airplanes must be made safer, more controllable at low speeds incident to taking off and landing—where most accidents occur—and less expensive in initial cost and in the cost of maintenance and operation.

Although the problems of rigid and semirigid airships have been seriously studied during the past two years, there is much to be learned about airships. In the present state of development of rigid airships, as exemplified by the remarkable flights of the Shenandoah from Lakehurst, N. J., to Seattle, Wash., and return, and of the Los Angeles from Friedrichshafen, Germany, to Lakehurst without stop, it is evident that material has reached the point where, with practice and experience, the possibilities in the field of commercial airship transportation can be determined.

The aircraft industry is in a poor condition, but the remedy is within the control of the manufacturers and the governmental agencies concerned. The committee has proposed a basis for solution of the existing difficulty.

In view of all the circumstances at present affecting the development and use of aeronautics in America and its relation to the public welfare and national defense, the National Advisory Committee for aeronautics submits the following:

## GENERAL RECOMMENDATIONS

1. *Scientific research.*—The continuous prosecution of scientific research on the fundamental problems of flight should be regarded as in the last analysis the most important subject in the whole field of aeronautical development, as substantial progress in aeronautics depends upon the continuous acquisition of knowledge which can be obtained only by long-continued and well-directed scientific research.

2. *Air Mail Service.*—The Air Mail Service should be continued under the Post Office Department and its ground equipment for night flying should be extended to cover the entire route between New York and San Francisco. When this is done, overnight transportation of mail by aircraft between strategic points, as for example, between New York and Chicago, should be pro-

vided at rates that will make such service eventually self-supporting.

3. *Commercial aviation.*—Rapid development of commercial aviation is primarily dependent upon increasing the reliability and economy of operation of aircraft. Other countries, notably England and France, have encouraged commercial aviation by direct subsidies, and their experience has indicated that unless governmental aid is given, directly or indirectly, commercial air transportation can not be financially successful in the present state of aviation development. Legislation providing for the reasonable regulation of aircraft, airdromes, and aviators, and affording necessary aids to air navigation along designated national airways would be most helpful. The establishment of landing fields generally would also stimulate improvement in the reliability and economy of aircraft operation and facilitate the development of commercial air transportation in this country on a sound basis.

4. *Military and naval aviation.*—There should be continued study of the air defense problem of the United States, and continued support of aviation development in the Army and Navy.

# INDEX